Algebra in Words

presents:

WORD PROBLEMS DECODED

Gregory P. Bullock, Ph.D.

Gregory P. Bullock, Ph.D. © 2015

Bullock, Gregory P.
Algebra in words presents: word problems decoded

ISBN-13: 978-1523302192

ISBN-10: 1523302194

MATHEMATICS/Algebra/General

STUDY AIDS/Study Guides

First Edition

The United States of America

This book is dedicated to making

The United States of America

#1

in education in the world.

CODE [noun]: A system of pre-designated words, symbols, sounds, and arrangements with assigned meanings to communicate a message.

DECODE [verb]: To translate, unscramble, or extract meaning from a coded message to convert it into a form which may be better understood.

Word Problems are the *Code*.

You are the *Decoder.*

TABLE OF CONTENTS

INTRODUCTION

Word problems may be one of the most feared and disliked parts of math. There is a major ongoing communication problem in the world of math education. The way that many textbooks and instructors *teach* doesn't connect with students, leaving too many confused, intimidated, and lost, which results in too many quitting and/or failing, who have the potential to do well. Not all students think and learn alike.

Many students have enough trouble with algebra. Many have more trouble with word problems because they involve new steps, such as setting up the equation, and other facets, such as using units. Yet, textbooks tuck word problems (or, applications, as they're sometimes called) into a few pages at the end of a chapter. Students are expected to do them, and word problems often carry more weight per question on tests, yet the textbooks do not place enough emphasis on, or even really *teach* word problems. Textbooks, and sometimes the instructors, often don't offer enough useful insight into how to do word problems leaving students and instructors stuck in a quagmire of trying to either "learn it the textbook way," or even worse, trying to wing it, often both to no avail.

This is why even more students dread *word problems* more than plain algebra, and understandably so. Many students don't know where to turn for clarity. Meeting with a tutor often results in the tutor doing the word problems for their client, and still, not much is learned. When test time comes, too often I see students trying to "do the math" to figure out if they can still pass if they skip all the word problems. That is a major problem... but it doesn't have to be this way.

Either word problems are taught as being *too much the same*, with their *differences* not accentuated, *or* taught as each being *different and random* instead of drawing attention to their *proper similarities*.

This book addresses those problems head-on as it is dedicated to the mini-subject of word problems, as it should be treated, by teaching the basics behind them as a whole, but also identifying key elements and accentuating the differences of each specific type.

Another problem is that textbooks tend to focus more on the *solving* aspect but not enough on the most important aspect: *The Setup*. Students usually don't have as much difficulty solving, but you can't get to the solving steps if you can't *set up* the equations.

This book introduces a new, simple, logical process for solving word problems from beginning to end, called The Bullock Identify/Template Method. The secret behind the setup is to properly *identify* the *type* of problem (which I teach) and using what I call "template equations" associated with each type. In this book, students can search the word problem bank by type or category, and from there, either use the template equation, or setup the equation from scratch, as I also teach.

I wrote *ALGEBRA IN WORDS: A Guide of Hints, Strategies and Simple Explanations* (2014) to bridge the communication gap from the technical math language to students who learn better by worded explanations. With its success, I wrote *ALGEBRA IN WORDS 2* (2015) which focused on more advanced topics using the *in words* explanations. Those books do not cover word problems because I wanted to focus specifically on helping students build a solid foundation in algebra. Then I wanted to take on word problems, focusing on them in a way that would allow students to learn them easier, faster, and more successfully. In the early stages of this book, I wanted to start from scratch and take a fresh look at them, trying to think like a new student learning word problems for the first time. I started testing my methods in my classroom to my students. Upon implementing my own methods and explanations, I saw students' (who followed my methods) success on word problems reach near 100% (with the exception of occasional errors). Before then, I was used to having to deduct massive points on tests due to failed or not-even-attempted word problems.

Until now.

This book reveals the simplicity buried under the veil of confusion and intimidation of word problems. There are 55 word problems in this book, all representative of the types typically learned in algebra classes (see the full list in Table of Contents under Annotated Examples). Each word problem is fully worked out and explained in language and logic that is easy to follow and understand. Realistically, word problems need more attention than they're typically given, meaning more, useful pages and explanations in textbooks, or better yet, their own book.

Now there is one.

This book will teach you everything you need to know about word problems, from the background, to interpreting, to identifying, to setting up, to solving and answering the questions, and even how to properly do unit conversions. It will put you in total control, leave nothing to be desired.

This is *WORD PROBLEMS DECODED*!

HOW TO USE THIS BOOK: The 4 Phases

This book is divided into 6 main sections, each of which focuses on a specific aspect of word problems, but are also all connected to help you navigate to the areas where you need the most attention. Studying with this book will make your study time shorter and more efficient, leaving you more time to focus on the *math*, and more time to sleep the night before a test. You are not expected to read this book from cover to cover as soon and as fast as you can. This book is designed to be used *as you go* through your class, *at your pace*. To maximize the benefits of this book, follow these 4 phases, which will guide you through the appropriate sections and topics *as you need them*:

Phase 1: Reading the basics and general background of word problems
Read from *this point* up through the following sections as soon as possible (about 30 pages):
- **Why Word Problems Matter**
- **The Concessions Contract**
- **The Word Codes (Definitions)**
- **Unknown vs. a Variable**
- **The Importance of the Equal Sign**
- **The Importance of Units**
- **Equalities, Ratios, & Conversions**
- **The Real Order of Operations: GEMA**
- **The Word Problem Procedure**
- **The Categories**

The next section is Detailed Explanations & Templates, which you may choose to skip for now (or, feel free to read it; it's about 60 pages). You can refer back to them when you get to each type of problem in class. And don't worry about trying to figure out how and when to access them. Each of the Annotated Examples has page references to their corresponding Detailed Explanation sections for if or when you want to see *how to set up the template equations*.

Phase 2: Do the Identify & Match Practice (pg 105)
This is a practice section that helps you work on and strengthen one of the most important steps in word problems: *Identifying the problem*. Equations can only be set up once you identify what you're working with. This phase and section will simply have you read random problems, identify key elements according to the *Identify* step, and then

choose and match it to the type of problem from the list of *Categories*. That's it. Directly after it, the Answers section will give the answers and give you a direct page reference to the fully worked out problems.

Phase 3: Use Alongside Homework
This book is a great resource for using alongside homework. It is useful in a number of ways, whichever ways suit your needs best. It really just comes down to following The Word Problem Procedure (pg 33), starting by reading each word problem from your homework, and Identifying and Matching it to the type of problem it is in Categories. The list of Types in Categories should coincide with the chronological order in which you learn them in class. You can also easily find the section you need in the Table of Contents under Annotated Examples (but grouped differently, here, by common topic). The page numbers will direct you to a fully annotated example, which will guide you through all aspects of an example problem.

You might choose to look at the *Unknowns* to help you set up *your* unknowns. You might choose to go right to the *template equation* and plug-in the given numbers from *your* homework problem, then solve. The annotated examples in this book will likely use the same type of equation and type of math involved in solving. For more background on each problem, follow the page number given in the Identify step for *detailed explanations* on that particular type of problem.

This is how the reference trail flows:
• Search the Type either in Categories or Table of Contents. Follow the page number to the Annotated Example.
• Within the example, in the Identify step, a page number will be provided to Detailed Explanations for more background on the equation for that problem.
• Within Detailed Explanations, page numbers are provide to help you easily flip back to the associated Example problem.

While doing your own problems, follow this course of direction:
• Read the problem.
• Write your unknowns.
• Turn to The Categories section (pg 39) of this book.
• Using clues in your problem, choose the category that best fits your problem. From there, go to the page of the related example.
• Look at the template equation and write it on your paper.
• Then, using your givens, fill in the template equation to set up the equation.

5

- If you need more background, follow the page number in the Identify section to go to the Detailed Explanations section of the book.
- Otherwise, solve the equation(s) you've set up.
- This book provides step-by-step instructions to solve, however, if you need more help with the algebra, use *ALGEBRA IN WORDS*.

If you use this book to study concurrently with your class, hopefully you're reading the background of each type of problem in Detailed Explanations. Knowing where the template equations come from will make memorizing them easier because you will see the simple logic behind how they are made.

Phase 4: Studying for Exams

In Phase 3, you used this book to help you through homework. That was a learning phase. Now, you will need to retain the essential pieces to use during exams when you won't have access to this book. To prepare for word problems on exams, here's what to do:
1. Focus *only on the types of problems you will expect*. Go through your textbook or syllabus schedule and…
2. Identify the types of word problems you've covered since last test. Each test is likely to have no more than about 4 types of word problems related to the type of math you've been doing. Search them in Categories, then…
3. Focus on naming the unknowns and
4. Memorizing the template equations.
5. For more information, read through the example problems.
6. It is highly recommended that you do as many practice exercises from your textbook as you can, using the techniques and following the similar examples in this book. Check your answers.
7. Prepare in enough time so if you have questions, you can ask your instructor. Don't be afraid to ask your instructor a question even the day of the exam before it is handed out. (My students do, and I answer them).
8. If you are allowed a "cheat sheet" for your test, write the names of the problem types and template equations on it.
9. If your tests are open-book, ask if you would be permitted to use this book.

What about studying for midterms and final exams? There will likely be word problems from a larger pool of possible problems. Again, if you use this book concurrently with your class and chapter tests, you will be *much more familiar* with the setup and template equations related to each type of problem. After using this book and doing enough word problems, you will be much more prepared for what to look for, you will be much more familiar with "where the template equations come from," and you will notice the similarities of the problems.

Additionally: If you come across a word with a definition or relevance you are unsure of, check The Code Words (Definitions), starting on page 13, as there is a good chance it is included with a full definition and explanation of how it applies to each associated problem. Please note that the eBook contains hyperlinks so topics, words, equations and examples can easily be jumped to in a single click.

WHY WORD PROBLEMS MATTER

For any students who ever asked, "When am I ever going to use algebra?"
they now have one answer: in Word Problems. But I know this isn't enough. Students are then inclined to ask, "When am I ever going to use word problems in life?"

The answers to that question may not be as direct as students wish, but then again, they might be. In *ALGEBRA IN WORDS*, I said that the reason for learning algebra is to get a glimpse into the order of the world and universe around us and that learning it makes us better problem solvers. That's something that can be applied to endless facets of life. Word problems have a little more to offer (than just algebra).

Word problems are the bridge between arbitrary math and the real world.

1. They help with reading comprehension, critical reading, and attention to detail, which are skills anyone can benefit from, in any field, in any job, and even in everyday life. How many times have you ever heard a question and thought "What are they asking?"
2. More importantly, they show that questions can be posed to solve real problems, in other words, asking the right questions to help get the right answers.
3. They show that questions can be answered with scientific, repeatable proof. They may even expose multiple routes to the same solution. But that's just the tip of the iceberg.

The problems you will do at first will be simple and gradually get more complicated. Imagine the questions experts ask to solve the world's problems in regards to:
- balancing budgets
- research and development
- medicine
- construction
- building a business
- computer technology
- energy efficiency
- social programs
- social media
- entertainment

- marketing
- taxes
- war strategies
- homeland security
- financial planning/wealth management
- space travel and rocket science
- etc.

These all start with questions… *word problems*.

That's good news. There's no doubt they can be challenging. But word problems exemplify the fact that something complicated can be simplified. And, that challenging questions can be answered definitively. This book proves that.

THE CONCESSIONS CONTRACT

Are you ready to conquer word problems? If so, you must make a full commitment to learn them. Textbooks have misled you. They devote so little attention to word problems that students think there is little to be learned. The truth is: You can't shortcut or underestimate the *learning* behind word problems. Only when you learn them fully will they then become easy. But, you *can* use shortcuts in the *setup* steps when you start to see the patterns and understand where the shortcuts come from. To achieve the full benefits of this book, you must concede and agree to the following terms:

☐ Follow: How to Use This Book to get the most out of this book with the proper timing and alignment with your syllabus/curriculum.

☐ Word Problems are not random. They all follow patterns and templates. You just have to know what to look for. This book will show you what to look for.

☐ Don't be discouraged by the amount of information used to explain these problems. This book provides a lot of information on word problems, both generally and specific to each type of problem. You must learn the full background to understand and do the problems. You have to do a little more now, but it will make the problems easier (and faster to solve) later. Part of my commitment to guiding you is showing all the intermediate steps, including how to build the template equations.

☐ Any word problem falls into a category, just as all algebra problems fall into categories specific to the principles used to solve them.

☐ Word problems usually involve more steps than regular math problems, mainly in the setup and the final steps. It's inevitable.

☐ They take up more room on your paper. For some problems, you are advised to start at the top left of a new page so you have plenty of room to write and map-out your steps, including any conversions, mini-calculations, and sketches.

☐ The numbers involved may be much smaller or larger than you are used to (in the hundreds, thousands, or millions), and may contain decimals. This is expected. Take full advantage of your calculator to do the arithmetic. Also, don't quit a problem because the numbers "look weird." Your answers may be correct. You can evaluate your solutions at the end.

□ Word problems are not as much about getting the right answer as they are about setting them up and solving them correctly. You're expected to solve problems *algebraically* and *to show all work*, not guess or work backwards using trial and error. If you solve without algebra, you are completely defeating the purpose. You not only won't learn what you're intended to learn, but you should also not expect to get full credit. You are likely to get more credit for a correct setup and wrong answer than for a correct answer without the correct setup or algebraic steps.

□ Some students think they discover shortcuts to problems (and they may), but sometimes they work by sheer coincidence. Beware: Applying the same "shortcut" may not work next time.

□ They might take longer to do on tests compared to non-word problems, so try to budget your time accordingly. However, this book will help you do them very quickly, saving you time for other problems.

□ You may need to incorporate formulas. Some may be given to you directly, some you may be expected to memorize, or some you may have the opportunity to look up; it depends on your instructor. Always be willing to look up a formula. If you think you need one but can't remember it, ask your instructor for it. It can't hurt to ask. Instructors often care more that you're trying to work towards the solution than whether you memorize a formula.

□ You must be conscious of all units involved. (Units are covered extensively in this book).

□ Those who write Word Problems probably aren't trying to trick you. In fact, they're trying to feed you the information you need to set up equations you are familiar with. Word Problems are meant to be solved with the math principles and solving techniques you most recently learned.

□ Sometimes it's helpful to not over-think it. This is why my methods (templates) are useful: You will have an idea what the equation(s) will look like, instead of trying to create equations completely from scratch (although, this books teaches you how to do that too).

☐ Approach each one with an open mind and a positive attitude. This book will help you discover that they're not as scary as they may seem.

☐ Since there are many steps, there are more potential places for human error (making stupid mistakes, as they're sometimes called). It's normal, and it happens to everyone, even the best of them. Don't get discouraged, and don't quit. Show as much work and logic as you can, so, even if your final answer is wrong, you may receive partial credit, and you can go back later to find your mistake.

☐ Space and clarity are key. Map out your process so your steps are traceable, not only for the instructor who is grading, but especially for yourself.

☐ Although this book extensively covers the types of problems you will have to do, it may not cover every single type. Don't be afraid to innovate.

☐ You can do it. It will all make sense soon.

☐ I'm ready to make the commitment to solve word problems.

I (sign here)_____ hereby concede and agree to the above mentioned terms herewith.

THE BASICS OF WORD PROBLEMS

THE CODE WORDS (DEFINITIONS)

The Components of Word Problems

Integer – Any positive or negative non-decimal (and non-fraction) number, including zero.

Unknown – An unknown is just an unknown value for which the problem needs to be solved for. An unknown may be a single variable, but all unknowns are not necessarily variables.

Unknowns may be *in-reference-to* a variable. In this case, such unknowns are defined (as, what I call, *mini-equations*) before setting up the equation associated with the word problem. The whole unknown (as a mini-expression) is plugged into the appropriate place in the main equation. The reference variable is then solved for. Then, the value of the reference variable must be substituted back into the mini-equations of any other unknowns to solve for their values. This is explained more in Unknown vs. a Variable (pg 24).

Reference Variable –When there is only one unknown in a problem, that unknown will always be represented by a variable. However, some word problems have more than one unknown but only have enough given information to make one equation (and therefore, you can only use one variable because you can only have as many equations as you have variables; otherwise, solving would be impossible). The variable which other unknowns are described in-reference-*to* is the *reference variable* (I also sometimes refer to this as the **root variable**).

Throughout this book, "x" will often be used as the *reference variable*, unless the formula used in the equation has pre-defined variables (such as in Rate of Speed, Investment/Loans, and Geometry formulas), in which case, you can use one of them.

The reference variable is always mentioned after "than" or "as." For example, if a statement says "the width is three less than twice the length," the length is mentioned *after* "than," therefore, length (as L) would be the reference variable. Or if, for example, a statement said "twice as many dimes as nickels," nickels (as n or x) would be the reference variable and number of dimes would be expressed as "2n".

13

Reference-to-a-variable – An unknown which is not a single variable but is *in-reference-to* a defined variable (the *reference variable*). These can come in two forms: with *compensation factors* or a *total*. Unknowns in-reference-to the variable using compensation factors will be used with key-reference-words such as "more than," "less than," "half," "double," "three times as," etc. These words are then translated into symbolic form (as numbers, variables, and arithmetic symbols), still in-reference-to the unknown described.

There are times when a *total* amount is given and can be used as a reference-to-the variable as:
(total − x) = other unknown. This is often seen in problems where there are two unknowns, and there may or may not be compensation factors mentioned. Using the *total in-reference-to a variable* is logically done knowing that the reference variable, x, plus the "total − x" (which you can think of as "the rest") equals the total amount. These are often used in Mixed Items Problems.

Reference-to-a-reference-to-the variable (a.k.a. **Reference-to-an intermediate unknown**) – Usually the third unknown, among multiple unknowns, which references the middle unknown (the reference-to-the variable), both in words and algebraically, in a problem where only one equation will be set up. During the *naming the unknowns* step, the middle (or *intermediate*) unknown can be assigned a new variable (as well as in-terms-of the root variable), so the variable for the intermediate unknown can be substituted into the third unknown which references it. Then, the mini-equation from the intermediate unknown in-terms-of the root variable should be substituted appropriately into the third unknown so the third unknown is also expressed in-terms-of the root variable, thereby making all three unknowns in-terms-of the same variable, making them like-terms, allowing the variable to be solved with one equation. Once the value of the variable is solved for, that value should be plugged back into the mini-equations of the other two unknowns in order to find their values. This is discussed more in Detailed Explanations and an example of this can be seen in WP6 (pg 115) involving three different value coins.

Compensation Factor – A number that is added or subtracted to a term or side of an equation, or multiplied or divided by a term, which acts to equalize or balance the values on both sides of the equation. These are usually *given* in word problems describing one variable or value or unknown in-reference-to another variable or value, to help setup the equation. See how these play a role in balancing equations in: The Importance of the Equal Sign (pg 25).

Mini-Equation or **Mini-Calculation** – An equation or related calculation which must be done either as a preliminary step to set up the main equation, or near the end of the problem after the main equation has been solved. I consider any unknowns in-reference-to the variable as mini-equations since they must be solved by plugging the value of the solved variable into them at the end. I also consider *conversions* as being mini-calculations since these must sometimes be done in the beginning or end of a problem so values are in proper, synchronized units.

Template Equation – This is a phrase I have coined which means an *equation-shell* specific to each type of word problem, of which certain elements of the same type of problem are the same from problem to problem, but where different given numbers can be *plugged-in* to the appropriate places. *Template equations* are a major key to unlocking the simplicity of word problems because they reveal a consistency that most textbooks don't reveal, and therefore most students don't know exist. These are close to what we know as *formulas*, but they aren't quite formulas because they are unofficial and don't have designated variables for each element (as real formulas do). But they are definitely *formula-esque* because they are constructed around the mathematical rules of units, follow consistent patterns, and are by no means a result of randomness.

The Problem vs. The Question

It's very important to use each of these words in the proper context to best communicate what is going on. The *problem* is not necessarily one in the same as the *question*. The *problem* is the entire written statement, including the givens and the questions. You might even include all the steps used to solve as part of "the problem." The *question* is simply what is being asked and what answer is being sought.

The Solution vs. The Value of the Variable

The question gives direction as to what the *unknowns* and the variable are. The *solutions* answer the specific questions. You may *solve the equation* to find the *value of the variable* in the problem, but that isn't necessarily the *solution*. You may need to use the value of a solved variable to solve for other unknowns to answer the specific questions. This is why it's good to clearly name the unknowns.

GCF – The abbreviation for Greatest Common Factor.

LCD – The abbreviation for Least Common Denominator.

Degree of an Equation – The highest sum of exponents of each individual term, compared to the degrees of all other terms, in an equation. Remember that there can only be a maximum number of solutions equal to the degree of the equation. There can also be fewer, or no solution.

1st Degree (Linear) Equation – An equation in which the highest degree of any term is 1, usually involving x (to the power of 1, which is unwritten). Although terms with x may appear more than once, they can be combined, via addition or subtraction, into one term. The next and last step is usually solving for the variable using multiplication (of the reciprocal of the coefficient in front of x) or division (by the coefficient in front of x). 1st degree equations, by definition, are linear equations (even if they are not in slope-intercept form). Remember that a 1st degree (linear) equation can only have a maximum of one solution.

Quadratic Equation – A 2nd degree equation in which the highest degree of any term is 2, usually as x^2.
Depending on the equation, these can be solved by:
- The Quadratic Formula
- Using the Square Root Property (taking the square root of both sides)
- Factor & Solve: Moving all terms to one side set equal to zero on the other side, Factoring, setting each factor equal to zero, and solving each factor. Factoring may be done by:
  - Factoring out a Greatest Common Factor
  - Trial & Error or the Reverse FOIL Method
  - The ac/Grouping Method

Rational Equation – An equation involving *rational expressions* (fractions with variables). You should always begin solving rational equations by determining the *extraneous solutions*. Then, proceed by determining the Least Common Denominator (LCD) and multiplying all terms by the LCD in order to eliminate all denominators and bring the variables to the numerator where they can be solved. Multiplying all terms by the LCD will often result in either a linear equation or a quadratic equation.

Extraneous Solution(s) – Values which would make the denominator in a rational expression (fraction with variables) equal zero, and would cause *no solution*, or the equation, function, or associated graph to *not exist* at points with those values. This is because *any number divided by zero yields no solution*. Extraneous solutions are found by setting any and all denominators containing variables equal to zero, then solving (for x). These values should be compared to the purported solutions of a problem. Any purported solution that was also found to be an extraneous solution must be thrown out as a solution to the problem. These are typical in rational equations.

Note: GCF, LCD, and The math behind: Quadratic Equations, Polynomials, Factoring and Multiplying the "Special Cases," Radicals, Rational Equations, Extraneous Solutions, and Solving Quadratic Equations using these methods are all covered in *Algebra in Words: A Guide of Hints, Strategies and Simple Explanations*.

Subscript – A number, variable, symbol, word, or even phrase written in small font to the lower right of another variable or symbol. Subscripts are commonly used as descriptors to the variable or symbol to which they are attached, especially when the variable or symbol is the same as other used in an equation, but represent different values or tenses. It is important to note that subscripts play no part in the calculation.

Ratio – A relationship of one value to another, often expressed in fraction form, placing a value and its corresponding unit in the numerator, and a related value and its corresponding unit in the denominator. The units may or may not be the same. A ratio represents a constant value by which an unknown value in either the numerator or denominator of another ration can be calculated by proportion.

Proportion – An equation which sets one ratio equal to another ratio, in which the units correspond, and in which the value of each ratio is the same, but the numbers may be different. Proportions are used to find the unknown value in either the numerator or denominator of one ratio, using the other ratio as the constant or standard. These can often begin to be solved by cross multiplication or by multiplying both sides by the LCD of both ratios. For more, see: Ratios & Proportions (pg 47), WP34 (pg 171) and WP42 (pg 187).

Rate – Any ratio of some unit in the numerator with respect to time in the denominator. In this book, the two typical forms of rate are *rate of speed* (pg 23, 69) and *(yearly) interest rate* (pg 56).

Geometry Words

Length – In rectangles, length is the vertical stretch, and sometimes length is the same as *height*. Length is also a generic term that refers to measure of distance. For instance, asking "how long is the width?" is asking for the distance on that particular side. Length is a one-dimensional measurement. It can be referred to as the first dimension.

Width – The horizontal stretch. It is worth noting that neither the length nor width connote that one is longer than the other.

Height, a.k.a. **Altitude** – The distance to the top of the shape by drawing a straight dotted line upwards at a right (90 degree) angle from the base up to the highest point of the shape. It answers "how high is the bottom to the top?" It is worth noting that this dotted line will not always *touch* or connect with the apex of the triangle (or shape); but it will connect to the invisible ceiling projected out from the highest point of the shape.

Sometimes **length, width, height**, and **depth** are used interchangeably in reference to a 3 dimensional object.

Diameter – The longest length from one point on a circle to another point on the circle. The diameter runs directly through the middle of a circle, cutting it into two equal halves.

Radius – Half the diameter of a circle. If the diameter is given, the radius can be found by dividing the diameter by 2.

Perimeter – The distance around a two-dimensional shape, also known as *the sum of lengths of the sides of a shape*. Since perimeter measures the *distance around*, the units will always be in one dimension. **Circumference** is a specific type of perimeter – the distance around a circle.

Area – A calculated measurement of the space within a two-dimensional shape, or of the surfaces of a three-dimensional shape (surface area). Area is expressed in two dimensional units, such as cm^2, $inch^2$, or "square inches." Measurements of floors, carpets, and lawns are typically given in square units of area.

Volume – A calculated measurement of how much space is taken up by a three-dimensional shape. Volumes of symmetrical 3D shapes can be found with certain measurements and formulae. Finding the volume of a non-symmetrical shape, such as a random rock, can be found by liquid water displacement which calculates the difference in water levels after completely submerging an object under water. Examples of 3D units are shown in The Importance of Units (pg 27) and WP44 (pg 191).

Similar Triangles – Triangles which have the same three angles and, likewise, have corresponding sides which are proportional in length. Similar triangles are often used in problems where the unknown length of a side can be determined using proportion. A common type of word problem using similar triangles and proportion refer to a pole or person and a cast shadow or a distance from a building as in WP42 (pg 187). It can be helpful to draw sketches for such problems.

Hypotenuse – The side of a triangle across from the right angle, also known as side "c" in the Pythagorean Theorem. The hypotenuse is always the longest side of a right triangle, without exception. The hypotenuse is referenced in WP40 (pg 183) and WP42 (pg 187).

Leg – A side of a triangle that is attached to the right angle. By definition, a leg is a side of a triangle across from an acute angle. Since a right triangle always has two acute angles, a right triangle has two legs. Legs can also be called *the two shorter sides of a right triangle* and are represented by "a" and "b" in the Pythagorean Theorem, as seen in WP40 (pg 183) & WP42 (pg 187).

Right Pyramid – A pyramid whose apex is directly above the center of the base, as referred to in WP44 (pg 191). If a line were drawn from the center of the base to the apex, it would be at a 90 degree (*right*) angle.

Money, Investment, & Business Words

Appreciation, or **Appreciates** – The value of something increases since the time of purchase.

Depreciation, or **Depreciates** – The value of something decreases since the time of purchase.

Interest – The money resulting from a certain principle amount of money invested at a certain yearly interest rate. In the investment sense, interest is the new money gained on the principle. In the loan sense, interest is the extra amount of money due back to the lender, in addition to the principle amount borrowed. This is represented by variable "I".

Simple Interest – The money made from one investment period or cycle. Simple interest is the result of a short-term, one cycle investment. This is based on the Simple Interest Formula (pg 56).

Interest – The money made over a number of investment cycles in which the interest gained from the last cycle is reinvested as the new principle. This is sometimes referred to as "interest on interest," and is usually in reference to long-term investments, resulting in exponential growth, as seen in WP55 (pg 220).

Principle – The original amount of money invested, borrowed, or lent. This is represented by variable "P". For more, see: Simple Interest Investments (pg 56).

(Yearly Interest) Rate – The percentage that a *principle* invested is expected or agreed upon to increase by the end of the investment (or loan) time period. Interest Rates are given as percentages in the word problems but must always be converted to decimal form when put into an equation. This is represented by variable "r".

Time Period, or just **Time** – Time, when used in regards to *yearly interest rate*, is given in years. Mathematically, it must be this way so that the unit *years* cancels with *years* from the "per year" part of the *yearly interest rate*. In many related word problems, the investment time period is less than a full year, so the *time* is given as the appropriate fraction or decimal of a year. The symbol for time is lowercase "t" (not capital T, which stands for temperature).

Loan – An amount of money borrowed from a lender which is expected to be paid back in full plus interest within the agreed upon time period, at a certain interest rate. Loans based on simple interest are very similar to investments of simple interest because they both follow the formula: I = Prt. The only difference is that, for loans, the principle amount borrowed plus interest is due back to the lender, whereas, for an investment, the investor gets back the principle amount plus the interest earned.

Unit Price – The cost or value in dollars per unit or item. A unit may come in the form of a coin, a ticket, a manufactured good, a pound or gram of something, as seen in Mixed Items problems, and a square foot, as seen in WP43 (pg 189).

Gross Profit, a.k.a. **Gross Margin**, **Gross Income**, and **Revenue** – The profit made due to the sale of goods or services, with the Cost of Goods deducted. This is also often referred to as the profit made before expenses are deducted.

Expenses – The amount of money spent on various aspects of running a business or providing goods and services.

Net Profit - The amount of profit made from the sale of goods or services after expenses are deducted from the gross profit.

Time Cycle Words (used w/ Investments & Loans)

The following words refer to *how many times per year*:
Annual or **annually**: Once per year, inserted as "1"

Biannual, a.k.a. **Semiannual**: Twice per year, inserted as "0.5"

Triannual: Three times per year, inserted as "0.33"

Quarterly: Four times per year, inserted as "0.25"

Beyond four times per year, we usually say "five times per year," etc. (even though there are words for more than four times per year).

The following set of words refers to *how many years in a cycle* before something occurs again. Notice these words have "ennial" in them, with an "e":

Biennially: Once every two years, inserted as "2"

Triennial: Once every three years, inserted as "3"

Quadrennial or **quadriennial**: Once every four years, inserted as "4"

Quinquennial: Once every 5 years or every fifth year.

Bicentennial: Once every 200 years.

Millennial: Once every thousand years;

… and there are many other words involving *years, months, weeks and days*, which can be seen with examples, in Detailed Explanations of Simple Interest Investments & Loans (pg 56).

Rate-of-Speed Words

Rate of Speed, a.k.a. **Velocity** – A ratio of distance over time: $\frac{\text{distance}}{\text{time}}$
Common specific examples are: $\frac{\text{miles}}{\text{hour}}$ or $\frac{\text{meters}}{\text{second}}$
and there are others. Often, the number in the numerator is divided by the number in the denominator to get a collective, standalone value written over (an unwritten) "1", keeping each unit in place. For instance, the density of ethanol can be given as
$\frac{4\text{ g}}{5\text{ mL}}$ or $\frac{0.8\text{g}}{1\text{ mL}}$

where the quotient of 4 divided by 5 is expressed as 0.8 in the numerator. See more on Rate in Detailed Explanations of Rate of Speed starting on page 69.

Current – The speed of water (as in a river) or of air (as in a jet-stream) through which a vehicle travels. A vehicle may travel in no current as in still-water (think of a lake) or still-air (no wind). However, if and when a vehicle travels in a current, this affects the speed of the vehicle.

If a vehicle is traveling *with* (in the same direction as) the current, it is said to be traveling **downstream**. The current *helps* the vehicle and results in the vehicle traveling *faster* than against a current or in no current.

If a vehicle is traveling *against* a current, such as up-river, towards the wind, or against a jet-stream, it is said to be travelling **upstream**. When a vehicle travels upstream, the current causes *resistance* against the traveling vehicle and results in the vehicle traveling *slower* than with the current or in no current. This is continued in:
Rate of Speed: Upstream/Downstream (pg 78).

Unknown vs. a Variable

A *variable* is a letter or symbol that represents a value or a yet-to-be-determined value. An *unknown* represents a yet-to-be-determined value as well. An unknown may be a variable or a mini-equation made of part variable and part number (usually a compensation factor or a total).

An unknown in-reference-to a variable using a compensation factor where "the length is three more than 5 times the width" would be expressed as:
Let L = 5W + 3
as seen in WP36 (pg 175).

An unknown where the variable is in-reference-to a *total* might look like:
Let (5000 – x) = the other portion of the $5000 invested
representing the *rest* of an amount of money invested, as seen in WP22 (pg 141).

An unknown may also be a mini-equation made of more than one variable, as in:
Let (x + y) = the rate of speed of a vehicle moving with the current, as seen in WP29 (pg 158).

The Importance of the Equal Sign

The *equal sign* expresses *equality* and *balance* in value from one side of an equation to the other. It's what defines an "equation," even if and when both sides look differently. *Adjustments* are made on one or both sides of the equals sign to keep the equality. These adjustments can be thought of as *compensation factors* to make up for a greater or lesser comparative value. These compensation factors can appear as arrangements of different operations (addition, subtraction, multiplication, division) with different numbers, variables, and references-to variables. This allows for two things:

- It allows adjustments to be made to reflect and maintain the equality *during equation setup*, and

- It is what makes it possible to simplify to *solve for a variable*, which is why you perform the *same action to both sides* of an equation to ultimately isolate the variable.

Compensation factors are given in the problems in-reference-to some other quantity. For instance, "two less than the length" is the instruction to write
"L − 2" to balance the value between the width and the length:
$W = L - 2$
This is said and written this way because the length is greater than the width, so by subtracting that compensation factor (from the larger quantity), their values balance out and can be set up on opposite sides of the equal sign. Likewise, the same thing could also be written as:
$W + 2 = L$
again, because adding 2 to the smaller quantity equalizes it with the larger quantity (in this example), the length. For instance, if the width is 5 cm and the length is 7 cm, subtracting 2 from 7 makes both sides equal; or, adding 2 to the width of 5 makes both sides equal.

Words do not always translate into the equation in the same chronological order they are written in the word problem. Look at the example on the last page. "Two less" is written *before* "the length" but the "- 2" must be written *after* "L" in symbolic form because it is two less *than* the length. Likewise, notice how
"three less than twice the unknown" would be written in symbolic form as "$2x - 3$".

You will notice this order matters for "less than" because *order matters* for subtraction. But for a quantity "more than" or "greater than" some other quantity, the order in which quantities are added doesn't really matter due to the *commutative property of addition*. That being said, the literal translation of "4 more than L" would still be "$L + 4$", and to ingrain consistent practices, should be written as such.

Sometimes the equal sign can be used to set two equations equal to each other when each equation has a common quantity regardless of whether the value of the quantity is known. When a quantity is said to be *the same* between two equations, but *the value is not given*, this is a clue that your setup will involve setting the two equations equal to each other, using any given compensation numbers. This can be seen in the setup of WP27 (pg 154) where the rate of speed of two vehicles is the same but unknown, WP30 (pg 161), also involving rate of speed, WP39 (pg 181) involving the areas of two squares, and WP41 (pg 185) involving the areas of two circles.

This can also be done when each equation has a common variable with same quantities, regardless of whether the value is known. When the value of a quantity is said to be *the same* between two equations, but *the value of that variable is not given*, this is a clue that your main equation will be built by rearranging each sub-equation by solving for their common variable, then setting those equations equal to each other. Examples of this can be seen in WP24 (pg 145) and WP28 (pg 156) involving the rearrangement of the Rate of Speed Formula.

The Importance of Units

Units are one of the foundations of word problems. Algebra and numbers are brought to reality through units. Just as quantities and values on the left and right side of the equal-sign must be equal, the units must be the same on both sides as well. This is why units are the foundation from which all equations are built.

This is the major difference between word problems and straight math. In straight math, all numbers and values are "arbitrary" (unit-less, or having no identity). However, for word problems, *all* numbers and values (and even the values of symbols whose values aren't known yet) carry units. This is one of the ways which makes word problems more relatable and realistic than plain math. It is also what makes them a little more complicated.

Here's some good news: If you're at all familiar with variables, you can easily deal with units because they are treated almost exactly the same. Here are 4 ways how:

1. Just as like-terms can only be combined with like-terms, the same is true about like-units: *Only quantities of like-units can be combined.*

2. Just as a common factor in the numerator and denominator (either of the same fraction or fractions being multiplied) can cancel-out: *A unit in the numerator can cancel-out with a like-unit in the denominator.*

3. Just as a variable that is multiplied times itself can be converted into exponential form: *When a quantity is multiplied by another quantity with the same units, the units can be expressed in exponential form* (explained below, for area and volume).

4. *When numbers with units are multiplied, divided, or have exponents or roots applied to them, their units are affected the same way as their associated numbers.*

The only main *difference* is that variables *represent* values and can be solved for, whereas units are *descriptors* of those values as you will see more in the Detailed Explanations section. Units can guide you in setting up an equation and in determining how to rearrange an equation.

The most basic equation is built upon:
unit = unit

Here's a basic equation exemplifying quantities of same units being added:
Like-unit + like-unit = total of like-unit

Think of the units of mass:
grams + grams = total grams

Likewise equations of same-units can involve compensation factors of the same-units, as in:
miles = miles +/- compensation number in miles

If different units of mass are given, such as:
grams + pounds = ?
they cannot be combined as they are. In order to be able to combine them, one unit or the other must be converted; this is covered in: Equalities, Ratios & Conversions (pg 30).

Let's take a look at how units are affected by a few other operations.

Take Area for instance, where the area of a rectangle is found by multiplying length times width:
$A_{rectangle} = LW$
Suppose the units of both measurements are in inches (in):
$(in)(in) = in^2$, pronounced as "inches squared" or "square inches."

Take the volume of a cube, found by length times width times height:
$V_{cube} = LWH$
where the units are in centimeters (cm):
$(cm)(cm)(cm) = cm^3$,
pronounced as "cubic centimeters," (or "cc's" as used by some in the medical field).

Continued on the next page…

What about when units which are not alike are multiplied ... think about the units for the equation of Force, where **F**orce is the product of **m**ass times **a**cceleration:

Force = (mass)(acceleration)

where mass carries units of grams (g) and acceleration carries units of "meters per second squared," $\left(\frac{m}{s^2}\right)$.

Multiplying the units of mass and acceleration yields $\frac{(g)(m)}{s^2}$

These units are repackaged as a unit of a Newton (N) which represents one $\frac{(g)(m)}{s^2}$,

which is why Force can go by units of N or $\frac{(g)(m)}{s^2}$.

Now look at a simple ratio, or division, of units. An easy one is rate of speed. As rate is literally $\text{rate} = \frac{\text{distance}}{\text{time}}$, you can see why speed may be given as miles per hour, $\frac{mi}{hr}$, because it is the result of dividing distance (in units of miles) by time (in units of hours). This is explained in more detail in: Rate of Speed (pg 69). Ratios can also be used as conversion factors, to convert from one unit to another, explained in the next section, Equalities, Ratios, & Conversions.

Equalities, Ratios, & Conversions

As explained in the last section, units must be alike in order to combine their numbers. In many of the word problems you're given, the units of numbers will be different. Part of setting up the equation or finishing the problem (by answering the question asked) will require unit conversions. You should understand where conversion factors come from and how to use them so you can successfully utilize them and show them in your work. This section will explain it to you.

The quantities or measurements of two things can be the same, but expressed with different values if the values have different units. These are called *equalities*. Equalities can be made into *ratios* and ratios can be used as *conversion factors*.

The best way to explain this is with an example comparing two different units of the same length, such as the equality:
1 inch = 2.54 centimeters

An equality can be put into ratio form by putting either value with its unit in the numerator and the other in the denominator. The equality above can be made into either conversion factor:
$$\frac{1 \text{ inch}}{2.54 \text{ cm}} \text{ or } \frac{2.54 \text{ cm}}{1 \text{ inch}}$$

Which one is right? Both are right, but the better question is: Which is the best one to use? The answer depends on what the goals are, such as: What units are being converted *from* (what are currently there) *to* what you need them to become? The answer to this will tell you which arrangement to use. For simple conversions, position the conversion factor (ratio) so the unit you're converting *from* is in the denominator and the unit you're converting *to* is in the numerator. As mentioned last section, treating units by the rules of variables, a unit in the numerator cancels-out with a like-unit in the denominator.

Continued on the next page...

Suppose we needed to convert 7.82 cm to inches. We would position it as:

$$(7.82 \text{ cm})\frac{1 \text{ inch}}{2.54 \text{ cm}}$$

Notice that the arrangement of the conversion factor placed "cm" (with its value) in the denominator because that is the original unit of the number being converted *from*, and inch (with its value) is placed in the numerator because that's the unit being converted *to*. "7.82 cm" can be thought of as a fraction by placing it over "1". To complete the problem, cross out "cm" in the top and bottom, multiply the numbers in the numerators to get 7.82, multiply the numbers in the denominators to get 2.54, then divide the product of the numerator by the product of the denominators to get 3.08 inch (rounded to the nearest hundredths place):

$$\frac{7.82 \; \cancel{\text{cm}}}{1}\left(\frac{1 \text{ inch}}{2.54 \; \cancel{\text{cm}}}\right) = 3.08 \text{ inch}$$

See examples where units of time are converted in WP23 (pg 143), WP24 (pg 145), and WP26 (pg 149).

The Real Order of Operations: GEMA

Remember, when simplifying expressions and equations and doing mathematical manipulations, *order matters*. Below is the true and complete order of operations. I recommend using the acronym GEMA (instead of PEMDAS), explained below.

1. **G**roups – Simplify groups first, if possible, from *inner to outer*. A group is a set of (parentheses), [brackets], {braces}, overall numerators, overall denominators, radicands, and absolute value groups. When speaking a mathematical statement, a group is often called "the quantity (then say what is in it)."

2. **E**xponents *or* roots, whichever come first, from left to right. Remember, any root can be converted into an exponent as a *rational exponent*. This step may include distributing an exponent to factors in a group or expanding a group raised to an exponent.*

3. **M**ultiplication *or* division, whichever comes first, from left to right. This step may include *distribution*.

4. **A**ddition *or* subtraction, whichever comes first, from left to right.

* Converting roots (radical form) to rational exponent form is covered in *Algebra in Words 2*.

THE WORD PROBLEM PROCEDURE

As you've read earlier, *word problems are not random*, neither in the problems given, nor in how they are grouped together, nor in how they are set up. There are two ways to begin word problems. Either by:
- Building the equation from scratch, or by
- My pseudo shortcut method (Identifying, then using Template equations).

Both approaches are explained in this book and will lead you to the same named unknowns, equation setup, and solutions.

The *building the equation from scratch* approach uses the fundamentals explained in: *The Importance of the Equal Sign* (pg 25), *The Importance of Units* (pg 27), and *Equalities, Ratios, & Conversions* (pg 30), and incorporates any *formulas* that pertain to the type of problem. These are all explained in the *Detailed Explanations* section which also shows how to build the template equations used in each type of problem.

My *shortcut method* involves: Identify the type of problem and plug-in to the Template Equation. It bypasses *building from scratch* by jumping right to the *template equation*. As far as I know, this method has never been done before, so I call it:
Bullock's Identify/Template Method.

Each word problem should have the following parts:
- **The Problem** (which is given)
- **Identify** (the type of problem)
- **Name the Unknowns**
- **The Template** (equation) **or Formula**
- **The Setup**
- **The Math**
- **Conversions**
- **The Solutions**

Each part is explained in detail, as follows…

The Problem
The problem will be given. Read the problem slowly and thoroughly the first time. Re-read it as many times as you need to; you can always look back at it. Then begin decoding...

Analyze the word problem for key elements and clues (to help you *identify the type* of problem) by looking for answers to these questions:
- What's the general topic?
- How many unknowns are there?
- What should you let the variable(s) represent?
- Are any unknowns *in-reference-to a variable*?
- How many variables will there need to be?
  - How many equations will there need to be?
- What values are given, and with what unit(s)?
- Are any *totals* given? If so, for which units?
- Are any quantities said to be equal (or same) even if the values aren't given?
- What exactly is being asked?
- Will any unit conversions need to be done either in the beginning or end?
- What type of equation(s) are involved?
- What kind of math will need to be used to solve?

You may not discover the answers to all these questions at first. If you don't, that's ok, especially for the last two questions, but now you know what to look for. Answer what you can, and use those clues for the next step...

Identify
As these problems are newer to you, let this book do some of the work. Use the clues you gleaned from analyzing the problem (using the questions above) to *identify* the type of problem it is from The Categories. Find the type that best fits your problem and follow the page number to a similar, fully annotated example. For more background, follow the page number in the Identify step of the example.

Sketch a diagram if possible, though, for many problems, it won't be necessary. It may be helpful to construct a diagram for problems involving geometry, to help conceptualize what goes where. Label it as detailed as possible, with names and with the variable(s) and unknowns. This may help you define the unknown(s)...

The Unknowns
Start this step by naming what the reference variable will represent. This should be easy. If it is obvious that there is one unknown, that will be what the variable x (or some variable) will represent. In problems which reference formulas, there are already predetermined variables you can use. This is likely to be seen in equations involving Rate of Speed, Interest, and Geometry. Use subscripts if necessary to distinguish between variables of the same letter but of different context. If there is more than one unknown, you will notice the *reference variable* will be mentioned after "than". For more on this, see: Reference-to-a-Variable (pg 14), Unknown vs. a Variable (pg 24), and The Importance of The Equal Sign (pg 25).

Write "let x = " (and write, *in words*, what x represents, including the proper units).
Follow the written clues in the word problem to write the other unknowns *in words* set equal to the mini-equations in-reference-to the variable.
If there can be more than one variable (such as "y"), name them as well. This will occur for Systems of Two Linear Equations and Systems of Three Linear Equations. The Detailed Explanations section will give you more guidance on naming the unknowns, specific to each type of problem, including template *unknowns* for problems with reoccurring patterns of naming the unknowns, so you can easily plug them into the equation and answer the questions at the end.

The Template
Based on the type of problem you've *Identified*, use the associated template equation. This is, in essence, the formula to use, unless there is an obvious, specific formula to use. What you will find in many cases is that the template equations are built from formulas.

Once you find the correct template equation, just plug the numbers in and solve.

Are you able to successfully fill in all the unknown aspects of the template equation? If so, proceed to The Setup. Otherwise, re-evaluate:
• Did you extract the right clues from the given problem to help you Identify? Try reanalyzing the problem by going through the suggested questions listed on the previous page.

- Is the unknown in a different place than shown in the template equation? If so, name your variables and unknowns properly and arrange them into the correct place(s).

- Maybe that exact type of problem is not covered in this book. In that case, look to the closest related problem for ideas and units, and use the techniques in this book to build the equation from scratch.

The Setup
Set up the problem based on the template you've been led to from the Identify step. Build your equation by plugging in the givens and unknowns. Problems involving Systems of Two Linear Equations will require setting up two equations.

Remember to do any necessary **mini-calculations** or **conversions** here so that all numbers fit in proper context into the equation. Convert percentages to decimal form and do any conversions so all units of numbers are synchronized. This may require you to do a mini-calculation off to the side. (Note: Units can also be converted *after* you've solved for the unknowns, to accommodate the details in the questions asked). For help on conversions, see: Equalities, Ratios & Conversions on page 30.

The Math
The Types of problems listed in the Categories section are grouped to chronologically coincide with your textbooks, and grouped in the Table of Contents by the type of math used to solve. The category heading and the *Detailed Explanation* may tell you the solving techniques you will use to solve. Be sure to show all steps in your work in case you need to trace it backwards. Remember, you should expect the math and solving techniques to be what you most recently covered in class. For instance, if you just covered *linear equations*, you will likely do word problems involving linear equations. If you just covered *quadratic equations, the quadratic formulas, the square root property, and radicals*, you will likely use those concepts to solve. Always be sure to follow Order of Operations and the proper procedures for solving algebraic equations.

Procedures for solving *The Math* to linear equations, systems of two linear equations, quadratic equations in all forms, rational equations, radical equations, factoring, using the LCD, simplifying radicals, and much more are all covered throughout *ALGEBRA IN WORDS* and *ALGEBRA IN WORDS 2*.

The Solutions

Once you solve for x (or any variable), plug the value into the mini-equations of any other unknowns. It's very important that you go back and *re-read the questions*. This will ensure that you don't overlook the actual questions being asked. Clearly state the solutions, *always including the proper units*. **Conversions** can be done independently from the main equations and can be kept separate from the equations. Before circling your answer, re-read the original problem and make sure you are answering the questions being asked.

Also, ask yourself if the answers seem logical. If they don't, trace back through your steps to search for an error, however, suppose you can't find an error, DON'T ERASE YOUR WORK. Often, students question their answers because "the numbers look weird," however, many times, they are still 100% correct. If you question your answers, never erase your work. Rather, put a slash through it and start over. This way, the instructor grading it can look at your first attempt and see that you tried, if not give you credit for parts that are correct (maybe the entire thing).

In addition to asking yourself if your solutions seem logical, it is a good idea to **check** your solutions by plugging them back in for the variable (or unknown) they represent in the original equation and simplifying. Upon complete simplification, the left side should equal the right side. If they don't equal, this may be due to one of three reasons:
1. The solution is invalid and must be discarded. This is common for problems involving quadratic equations where one solution of x will prove to be mathematically invalid.
2. There is *no solution* and you did everything correctly.
3. You may have made a mistake and should reevaluate your work.
The *check* steps are not shown in this book (because they are all verified to be correct, and) because the focus of this book is Identifying and Setting up the equations. However, checking is considered an important step.

THE CATEGORIES

Below is a list of the types of word problems in the **CHRONOLOGICAL ORDER** you are likely to encounter them. Follow the page number to the example problem, and follow from there.

1st Degree Equations using Proportion & LCD to Solve

Quadratic Equations using Factoring, the Quadratic Formula, the Pythagorean Theorem or Square Root Property to Solve

Exponential Functions

Extras

DETAILED EXPLANATIONS & TEMPLATES

Most of the following detailed explanations explaining the background on types of problems and their specific template equations are based off The Importance of The Equal Sign (pg 25) and The Importance of Units (pg 27).

The templates are like fingerprints for each type of problem. Template equations are like formulas specific to each type of problem. As is the case with any formula, there are segments (variables) which can be replaced with values for those variables. Since word problems don't have official *formulas*, they might not have official variables to fill in with values, which is why I refer to their *segments* which can be filled in with given values. An equation for a word problem might have official variables if the equation is built from a formula. In either case, templates make doing word problems much easier because, once you identify the type of problem, you can use the template equation as a starting point instead of having to create each equation from scratch.

Also, templates for certain types of problems will include *template unknowns*, which are patterns for how to name the unknowns. As with the equations, the named unknowns will likely be very similar amongst similar types of problems. The template unknowns give you a starting point which will allow you to fill in the proper segments with the appropriate words and mini-equations (for naming) and given values.

The templates are here for your reference. The intention is to *read* the problem, *identify* the *type* of problem, then use the template to set up the equation. If you need more background on the type of problem, it is provided here, in Detailed Explanations.

One Unknown, One Variable, One Equation

Problems like these can be recognized when the question only asks about one value. The variable will represent that only unknown. These types of problems can come in a number of forms. There may be one instance of x in the equation, there may be multiple instances of x, and x may be in the numerator or denominator.

Problems like these will use either a simple equation or a simple formula and are 1st degree, linear equations. The math will involve few steps, including combining like-terms and simple rearrangement techniques to isolate the variable. Some examples are an Unknown Among an Average (pg 109) and problems involving percent calculations as in WP13 (pg 128), WP14 (pg 129), and WP15 (pg 131).

One Variable, Multiple Unknowns, One Equation

Problems like these will become apparent once a reference to another unknown appears in the problem. One unknown will always be in the form of a variable, however, any other unknown will be in-reference-to the variable and will come in two forms:
1. with compensation factors, and
2. with respect to the given total.

Be sure to name all unknowns before setting up the equation. Any unknowns besides the variable should be named with mini-equations, as you will use these at the end of a problem by substituting the value of the solved variable into them to find their values. Naming the unknowns *first* will also help you set up the equation.

When *two totals* are given in the problem because it can be solved either with:
• one variable, unknowns in-reference-to the variable, and one equation, *or*
• two variables and two equations as a system of linear equations.
A sub-category of one variable problems are those with...

A Reference-to-a-Reference-to-the Variable

There are also more complicated problems where an unknown will *not* be in-*direct*-reference-to the root variable, but may be in-reference-to an *intermediate* variable which *is* in-direct-reference-to the root variable. Suppose it is given that Unknown 2 is one more than Unknown 1 and Unknown 3 is four less than three times Unknown 2 (in a real problem, these unknowns will have different labels). It will follow this general scheme:

Unknown 1: x is the root reference variable
Unknown 2: y is in terms of x as $y = x + 1$
Unknown 3: z is in terms of y as $z = 3y - 4$
Substitute "x + 1" in from Unknown 2 in for y in Unknown 3 to get:
Unknown 3: $z = 3(x + 1) - 4$

In order to do the problem, all unknowns must be *in terms of the same variable*, specifically the root variable, which, in this example, is x. Think of Unknown 2 as an intermediate between Unknowns 1 and 3. Before setting up the problem, you will have to do a mini-substitution, substituting Unknown 2 ("x + 1") in for y into Unknown 3 (as seen above).

When doing problems like these, start by determining and naming the *root variable* and work outward from there. Then name the intermediate unknown, then name the last unknown, then substitute the intermediate unknown into the last unknown so all unknowns are in terms of the same variable. This type of logic is used in WP6: Three Coins of Different Value w/ One Variable (pg 115).

Consecutive Integers

Consecutive Integers are numbers that follow in a sequence, one after the other. They may also come with an added condition such as being consecutive odd or even integers, which would count *every other* integer, as adding 2 to an odd integer yields the next odd number, and adding 2 to an even integer yields the next even number. Be on the lookout for other variations such as consecutive multiples of 3 in which each number in the sequence will be 3 more than its predecessor.

Note: Often, the variable "n" is used as the unknown variable when representing a *number*, or a number *in a series*, or a number representing "how many?" such as people or items. In *consecutive number* problems, n represents "a number." The truth is, "x" can just as well be used as the unknown variable, but x is more often used as a more general unknown variable or for independent data in which "y" values are associated, such as in functions or graphical relationships. Textbooks usually use "n" for these.

Notice below that consecutive even or odd integer problems are set up exactly the same. This is like saying "every other number, starting with the first number (n) in the series." All unknowns except for n are references-to-variable n. There is not much to the background of these template equations. The first step in the solving process is explained below to show how the equation should look. Complete the problems by substituting your solved value for n into each subsequent unknown mini-equation in-reference-to n.

Template Unknowns
Let n = the first integer
Let $(n + 1)$ = the 2nd consecutive integer
Let $(n + 2)$ = the 3rd consecutive integer
Etc.

Template Equation:
$n + (n + 1) + (n + 2)... $ = the sum of the consecutive integers
Add on as many additional consecutive groups as needed, increasing by 1 more than the previous group.

Problems like these are solved in a few simple steps. Combine the like-terms on the left which will form
$3n + 3$ = the sum #

for problems with three consecutive integers. Continue to simplify to solve for "n" then substitute the value of n into the mini-equations to find the values of the other unknowns. Back to WP2 (pg 110).

Consecutive Even *or* Odd Integers

Template Unknowns
Let n = the first integer
Let (n + 2) = the 2nd consecutive integer
Let (n + 4) = the 3rd consecutive integer
Let (n + 6) = the 4th consecutive integer
Etc.

Template Equation:
n + (n + 2) + (n + 4)… = the sum of the consecutive even or odd integers
Add on as many additional consecutive groups as needed, increasing by 2 more than the previous group.

Problems like these are solved in a few simple steps. Combine the like-terms on the left which will form
3n + 6 = the sum #
for problems with three consecutive even or odd integers. Continue to simplify to solve for "n" then substitute the value of n into the mini-equations to find the values of the other unknowns. Back to WP3: Consecutive Odd Integers (pg 111). Back to WP4: Consecutive Even Integers (pg 112).

Some problems involve consecutive integers but are not related by the sum of the integers. Instead, they set one integer equal to another using compensation factors to equalize their values.

Ratios & Proportions

In most related problems, you will be given one ratio which is to be used to find an unknown in a proportional ratio. The *given* ratio is the model or *standard* from which the other ratio will follow. This is why the subscript "standard" is attached to one of each unit in the equations shown below. There are two ways to set up a proportion. Each way is based on a consistent association of one unit to another.

Unit A to Unit B can be expressed as:

$$\frac{\text{Unit A}_{\text{standard}}}{\text{Unit B}_{\text{standard}}} = \frac{\text{Unit A}}{\text{Unit B}}$$

where the standards are in the same ratio and the same units occupy the same part (numerator or denominator) of opposite fractions, or:

$$\frac{\text{Unit A}_{\text{standard}}}{\text{Unit A}} = \frac{\text{Unit B}_{\text{standard}}}{\text{Unit B}}$$

where the same units occupy the same fraction, and the standards occupy the same part (numerator or denominator) of opposite fractions.

Problems which set up proportion equations are often positioned to use the LCD to begin to simplify and solve, however cross-multiplying can usually be used to start as well, setting up the same equation in the next step in either case. Depending on what variables are in the numerator and denominator to start, proportion equations may turn into quadratic equations. Back to WP34 (pg 171). Proportions can also be used to set up and solve percentage problems…

Percentage

All percentage problems are based on the formula:
(% in Decimal Form) (Original Amount) = Part of Original Amount

It is often insinuated that the number representing the *part* of the original is smaller than the original. That is true when the percentage is less than 100. Saying "**the original amount**" is a more inclusive phrase than calling it "the whole," even though they are essentially interchangeable in the proper context. Using "the original amount" leaves the door open for when the "part" is larger than the original amount, as is the case when the percent is higher than 100. In either case, the whole or original amount is the reference amount.

A typical question involving percent is "What percent of some original amount equals the part of that original amount?" where two of the elements are given and one element is unknown. If you were to do a literal translation from sentence to equation:
"is" means " = ",
"of" means multiply, and
"what number" is the unknown assigned as "x".

Percentage equations can also be set up as a proportion in which one ratio has the *part* in the numerator and the whole (or original) in the denominator, set equal to the *percent* (*not converted* to decimal form) in the numerator over *100* in the denominator:

$$\frac{part}{whole} = \frac{percent}{100}$$

This is sometimes also shown as:

$$\frac{is}{of} = \frac{percent}{100}$$

Then, cross multiply which would give:
(percent)(whole) = part(100)
and continue to simplify to solve.

Note: Be sure to convert any single digit percentages into their proper decimal form such as 5% to 0.05 (not 0.5).

All percentage **template equations** are based on:
(Original Amount)(% in Decimal Form) = Part of Original Amount
and depends on what the unknown is.

Take notice that we typically write the numeric coefficient before the variable, so the "x" may *switch positions* depending on if "original amount" or "percent in decimal form" is given or unknown. The commutative property of multiplication allows this, however, it is pointed out in case it looks differently than expected. This position switch can be seen in WP15 (pg 131).

When the *Part* of the Original is Unknown:
Let x = the part of the original amount
and use:
(percent in decimal form)(original amount) = x
or
(0.##)(#) = x

Or, as a proportion:
$$\frac{x}{\text{original amount}} = \frac{\text{percent}}{100}$$

Back to WP13 (pg 128).

When the *Original Amount* is Unknown:
Let x = the original amount
and use:
(percent in decimal form)x = part of the original
or
0.##x = #

Or, as a proportion:
$$\frac{\text{part}}{x} = \frac{\text{percent}}{100}$$

Back to WP14 (pg 129).

When the *Percent* is Unknown:
Here, x represents the percent in decimal form, in:
(original amount)x = part of the original
or
#x = #

Notice the "x" is written after the number representing "original amount." This is so the given value for "original amount" can be written as a numeric coefficient attached to the variable. Continued...

49

Once the value for x is found, it needs to be multiplied times 100 to convert it to *percent* form.

Or, as a proportion: $\dfrac{\text{part}}{\text{original amount}} = \dfrac{x}{100}$

Note: In the proportion form equation for percentage, the percent is *not* in converted decimal form. If the percent is 37%, "37" goes in. The decimal form could go in the numerator if the denominator is "1".

Note: The "part" of the original amount may be larger than the original amount if the percentage is greater than 100%. Back to WP15 (pg 131).

PERCENT INCREASE: Price or Salary

Any problem involving percent increase follows the equation:
(original amount) + (% in dec. form)(original amount) = new, higher amount

where the product of (% in dec. form) times (original amount) is the "amount of the increase":
original amount + amount from increase = new, higher amount

Percent Increase: *New* Amount Unknown

When the *new higher amount is the unknown*, the **template equation** will be: (original amount) + 0.##(original amount) = x

In this case, you can just solve, and the answer is the unknown. For instance, think about spending one hundred dollars...
100 plus 7% tax is calculated as:
100.00 + (0.07)(100.00)
100.00 + 7.00 = $107.00
because 7% of 100 is $7.00

It can also be done as (100.00)(1.07) = $107.00
(Think of 100% + 7% = 107%, and 1.07 in decimal form).

To look at the reverse (a 7% decrease), see: Percent Decrease (pg 52). Back to WP16 (pg 133).

Percent Increase: *Original*/Amount Unknown

The original amount, which is set to increase, can come in a number of forms. It may be a previous salary or may be the pre-sales-tax price of a good, as applying sales-tax will cause the final cost paid to *increase* by a certain percentage.

When the *original amount or salary* is unknown, add the % increase to 100%. In decimal form, that would be like 1.##. In other words, if the % increase is 7%, and added to 100% (representing the entire original amount), it would be 107%, and in decimal form would be 1.07. This is why either **template equation** can be used:
$x + 0.\#\#x = \#$

or: $1.\#\#x = \#$, where the two like-x-terms are added in the beginning.

Back to WP17 (pg 134).

Percent Increase: *Percent*Unknown

When the *percent is the unknown*, the equation is:
(original #) + (original #)x = #

where the "original #" is the same number in both spots.

Let x = the percent increase in decimal form
Into the equation above, substitute the given numbers in for original amount and the new, higher amount, and x for the unknown percent as a decimal to get the **template equation:**
$\# + \#x = \#$

Once you find the value of x, multiply it by 100 to convert it to percent form. Back to WP18 (pg 136).

Percent Decrease

From a conceptual standpoint, consider a question which asked:
What would the resulting amount be from a 7% decrease of 107.00?
Begin by taking 7% (as 0.07) of $107.00, then subtracting it from
$107.00:
107 – (0.07)(107.00)

107 – 7.49 = $99.51

The point here is to remind you that, although a 7% *increase* to 100 is
100 + 7 = 107, a 7% *decrease* to 107 is not the same as subtracting 7
because 7% of 107 is $7.49 (not $7).

Just as in the Percent Increase problems and examples, any part of the
equation could be unknown. Percent Decrease equations are modeled
very closely to percent increase equations, the only difference is the
minus sign between "original or *previous* amount" and the product of the
percent in decimal form times the original or previous amount:

(original amount) – (percent in decimal form)(original amount) = the
new, decreased amount

Let x = the unknown, whichever it is, and substitute it and the given
numbers into the equation above to get one of the following:

If the *new, decreased amount is the unknown*, the **template equation**
will be:
- (0.##)(#) = x

If the *percent in decimal form is the unknown*, the **template equation**
will be:
- #x =

If the *original, higher amount is the unknown*, as in WP19, the **template
equation** will be:
x – 0.##x = #

Back to WP19 (pg 137).

MONEY RELATED

Fees & Membership Costs = Total Bill

This type of problem may feature an initial fee plus additional costs based on proportional use. An initial fee may come in the form of a membership, start-up, or one-time fee. Due to the variability of these types of problems, the template equation for this type may not be exact, but it can generally be followed. Usually, the number of cycles is the unknown and the variable. Cycles may come in the form of minutes for phone use; megabytes for internet data; days, nights, weeks, weekdays or weekends, months or years for rental fees. Look for certain segments that require you to do a mini-calculation to put into the setup equation.

The equation is built around the units:
dollars + dollars + dollars = total dollars

You may have to do a mini-equation using given information to put into the main equation:
initial fee + (# of units)(cost per unit) + other usage or overage charges = total bill

Substitute the given numbers and x for the unknown number of units of time (whether minutes, hours, days, weeks, etc.) into the **template equation**:
$$\#.\#\# + (\#.\#\#)(\#) + \#.\#\#x = \#.\#\#$$

or some variation of this using the applicable segments. There's also the chance that the variable will need to appear in more than one place.
Back to WP10 (pg 123).

Expenses & Profit

All problems involving expenses and profit are built from of the formula:
Gross Profit – Expenses = Net Profit

where the units involved are:
Dollars – Dollars = Remaining Dollars

Let x = the time cycle unit, whether minutes, hours, days, weeks, etc. Some problems are simpler than others. A simpler version may look at two elements: one regarding gross profit and one regarding expenses, as in:

(Gross Profit)x – (Expenses)x = Target Net Profit

These can be specific to certain time cycles. The time cycles of the gross profit and expenses must (ultimately) be of the same unit. For instance, in this example, gross profits and expenses are on a weekly basis, and the unit of x, for time, will be "weeks":

(Weekly Gross Profit)x – (Weekly Expenses)x = Target Net Profit in x weeks

The #s being plugged into the **template equation** below could be of any size, from small to large, with few or many digits, and may include decimals:
#x - #x = #

Back to WP11 (pg 124).

Expenses & Profit (more complicated)

Some problems involving expenses and profits may be more complicated. They contain more elements of profit and expenses, and may be based on different units of time. These may also require mini-calculations to be done to set up the equation. These will still be built around:

Gross Profit – Expenses = Net Profit

where the units ultimately are:
Dollars – Dollars = Dollars

or more specifically,
(Gross Profit)x – (Expenses)x = Target Net Profit

In cases like WP12, *mini-equations* may be needed to determine the Weekly Gross Profit, the Total Weekly Expenses, and the Net Profit needed in a year (52 weeks) to fill in the equation above. The mini-equation for Weekly Gross Profit can be:

$$\text{Weekly Gross Profit} = \left(\frac{\text{\# tickets}}{\text{day}}\right)\left(\frac{\$}{\text{ticket}}\right)\left(\frac{\text{days}}{\text{week}}\right)$$

Notice above that "tickets" cancels with "ticket" and "day" cancels with "days" leaving the remaining units as "dollars per week" or "weekly gross profit." The mini-equation for Total Weekly Expenses can be:

$$\text{Total Weekly Expenses} = \left(\frac{\text{expenses}}{\text{day}}\right)\left(\frac{\text{days}}{\text{week}}\right) + \left(\frac{\text{other expenses}}{\text{week}}\right)$$

Notice above that "day" cancels with "days", leaving the units as "expenses per week" or "weekly expenses" which creates a like-fraction that can be added with the "other weekly expenses" fraction, giving units of dollars per week. Finally, since the goal net profit is based on multiple individual's current salaries, the following mini-equation is needed to calculate Net Profit Needed in 1 year (a.k.a. 52 weeks):

$$\left(\frac{\text{salary}}{\text{member}}\right)(\text{\# of members}) = \text{Net Profit Needed in 1 year}$$

where the units of "member" cancels with "members" leaving a salary in units of dollars. Letting x = the unit of time, the **template equation** will ultimately be something like:

#x - #x = #   Back to WP12 (pg 125).

Simple Interest Investments (or Loans)

The formula for simple interest is:
$I = Prt$

in which:
I is Interest
P is Principle (amount of money invested or borrowed)
r is (yearly) interest *rate*, given as a percentage but converted and applied in decimal form
t is *time* period, in *years* or fraction of a year.

The formula can be rearranged and solved for any particular variable.
Solved for P:
$$P = \frac{I}{rt}$$

Solved for r:
$$r = \frac{I}{Pt}$$

Solved for t:
$$t = \frac{I}{Pr}$$

Below are two common time cycle words expressed in years in the equation:
$I = Pr(\# \text{ of years})$

Biennially:
$I = Pr(2)$

Quinquennially:
$I = Pr(5)$

The unit of *time* should always be synchronous with the time-unit in the interest rate. For instance, if the interest rate is the *yearly* interest rate, which it almost always is, time should be in units of years. If time is not given in years, you must ***convert** all units of time into years*. If time is given in months, convert by multiplying given months by this **conversion factor**:
$$\frac{1 \text{ year}}{12 \text{ months}}$$

This does not mean a cycle can only be in *whole* years; a cycle may be in parts of a year, expressed as a fraction or decimal. If the amount of the year is given as a fraction, put the fraction in for t. If the time is given as a word, use these Time Cycle definitions (and the ones on page 22) to make the proper conversion. Below is a list of examples showing a word and the fractional part of a year it represents in the equation:

Biannually, a.k.a. **Semiannually**, which is *twice per year*, and the same as *every 6 months*:
$$I = Pr\frac{6}{12} \text{ or } I = Pr\frac{1}{2} \text{ or } I = Pr\,(0.5)$$

Triannually, which is *3 times per year*, and the same as once *every 4 months*:
$$I = Pr\frac{4}{12} \text{ or } I = Pr\frac{1}{3} \text{ or } I = Pr\,(0.33)$$

Quarterly, which is the same as *4 times per year*, and the same as once *every 3 months*:
$$I = Pr\frac{3}{12} \text{ or } I = Pr\frac{1}{4} \text{ or } I = Pr\,(0.25)$$

Quarterly (as in 13 weeks out of 52 weeks):
$$I = Pr\frac{13}{52} \text{ or } I = Pr\,(0.25)$$

If time is given as months, the fraction of time (still with the unit of "year") will be inserted as seen in the formula below:
$$I = Pr\left(\frac{\text{months in the period}}{12 \text{ months}}\right)$$

A period of **Monthly**, or 12 times per year, will appear as:
$$I = Pr\frac{1}{12} \text{ or } I = Pr\,(0.0833)$$

for each cycle. The same applies for weekly, daily, or by specific number of days.

Weekly:

$$I = Pr\frac{1}{52} \text{ or } I = Pr\ (0.0192)$$

Daily:

$$I = Pr\frac{1}{365} \text{ or } I = Pr\ (0.00274)$$

Specific Number of Days such as every 45 days:

$$I = Pr\frac{45}{365} \text{ or } I = Pr\ (0.123)$$

For problems involving a specific number of days through the course of certain months, use this paraphrase of **Mother Goose's Nursery Rhyme Mnemonic Device** to help you remember the number of days in each particular month:

Thirty days hath September, April, June, and November. All the rest have thirty-one, except for February, (which has 28, or 29 on leap years). Back to WP20 (pg 139).

Money can be split and invested in multiple investments at different interest rates. You are likely to encounter problems in which a total principle is split into two different simple interest investments (or loans) according to the equation:

$I = (\%)(x)t + (\%)(\text{total amount invested} - x)(t)$

In these types of equations, time is almost always "1" (so t is not shown) making the **template equation**:

$0.\#\#x + 0.\#\#(\# - x) = \#$

Because this deals with two unknowns, this is covered in different sections with similar equation setups, often grouped with *mixture problems*, either using one variable as in WP22 (pg 141), or two variables as in WP49 (pg 201).

It's important to be able to differentiate between "Interest" and "The new amount of total money." Interest is *only* the amount earned, which will always be less than the Principle (unless the rate was 100% or higher). The new amount of money is the sum of the original principle plus the newly earned interest:

new Amount of money = Principle + Interest

This is important to know so you can answer a question properly. If the question is:

How much interest was earned at the given rate in the given time period?

the answer is just the amount earned. If the question is:

After investing the given principle at the given rate in the given time period, how much is the portfolio now worth?

you must first calculate to find interest, then add that amount to the original principle, and that sum is the answer. This is explained with the formula in the next section: Simple Interest where the New Amount is Unknown (pg 60). Back to WP20 (pg 139).

Simple Interest: New Amount is Unknown

This can be done using either formula:
$A = P + I$
or
$A = P + Prt$

using the same variables from the Simple Interest formula, but now P appears twice, and A represents the new Amount. In this equation, "Prt" is substituted in for "I" and (its value) added to P, to show that:

Original Principle + Earned Interest = New Amount

Sometimes the formula:
$A = P(1 + rt)$
is given and used, but this is just an arrangement of the above equation, factoring P out as the GCF. If you distribute the P through the parentheses, you will again have:
$A = P + Prt$

Substitute the given values for P, r, and t into the formula, then simplify, and you will find the value of the new Amount, A. Back to WP21 (pg 140).

See Also: Compounding Interest in WP55 (pg 220).

GEOMETRY

In geometry problems, template equations aren't necessarily built the same way as they are for other word problems. Building geometry equations requires:
- finding the right formula (for the shape and unit of measure),
- naming the unknowns properly
  - using mini-equations and
  - compensation factors where necessary,
- and substituting the unknowns into the formula properly.

Still, though, template equations are given in this book, as each type of problem still follow common patterns. Like any other multi-unknown problem, you will have to plug the value of the solved variable into the mini-equations of the other unknowns to solve them. Problems involving a formula to solve for one unknown variable (with no other unknowns) are not covered in this book, as they just require basic algebraic rearrangement. This book focuses on geometry problems with more than one unknown.

Perimeter of a Rectangle

Problems involving perimeter of a rectangle will likely have both the length and width unknown. Clues will be given telling what the reference variable is and what the other unknown is in-reference-to the variable. The value of the perimeter will be given, allowing you to use the formula for perimeter of a rectangle:

$P_{rectangle} = 2L + 2W$ or $2(L + W)$

After naming the unknowns, substitute them into the perimeter formula. If L is the variable and W is in-reference-to L, the **template equation** will be like:

$2L + 2(\#L +/- \#) = P$

where "#" represent given compensation numbers. Or, if W is the variable and L is in-reference to-W, the **template equation** will be like:

$2W + 2(\#W +/- \#) = P$

Proceed by distributing the 2 through the parentheses, then combine like-terms and continue to simplify the linear equation to solve for the variable. Plug the value of the solved variable into the mini-equation of the other unknown to find its value. Finally, remember to report answers with the proper unit. Back to WP35 (pg 173).

Triangle: Sum of Sides (Perimeter) or Sum of Angles

Problems involving the perimeter of a triangle are very similar to problems involving the perimeter of a rectangle except that there will be three unknowns instead of two. Use the formula:

$$P_{triangle} = L_{side\ one} + L_{side\ two} + L_{side\ three}$$

There can be two variations of problems involving perimeter of a triangle:
- A simpler type in which two sides are both in-reference-to the variable (in their own given way),
- or more complicated problems that involve a reference-to-the variable, *and* a reference-to-a-reference-to the variable (see Note below). In either case, start by determining which side will be the reference variable, and name the other two unknowns according to the given clues. Substitute the unknowns and given value for area into the **template equation**:

$$x + (x +/- \#) + (\#x) = \#$$

Proceed by combining like-terms and solving for the variable. Then, substitute the value of the solved variable into the mini-equations of the other unknowns to find their values.

Just as a perimeter of a triangle problem involves the sum of all three sides, problems involving **the sum of the internal angles of a triangle** are similar. The only difference is that the value of perimeter (of any shape) has to be given whereas the sum of the internal angles of a triangle will never be given because they are always understood to be 180 degrees.

A problem involving the sum of internal angles of a triangle is not shown in this book but the logic involving naming the unknowns, setting up the equation, and solving are essentially the same as for perimeter. The equation used is:

$$Angle_1 + Angle_2 + Angle_3 = 180$$

The sum of internal angles of *any shape* can be found using the formula:
$\#$ of degrees = ("# of points" - 2)(180).

Note: The **more complicated type** of problem involving perimeter or sum of angles of a triangle where one of the three unknowns is in reference-to-a-reference-to the variable is not shown in this book, however WP6 (pg 115) closely mimics one, where the logic, naming of unknowns, and solving process are the same. Back to WP36 (pg 175).

Area of a Rectangle

Problems involving the area of a rectangle are very similar to those involving perimeter of a rectangle because they both involve length and width, both of which will be unknowns with one in-reference-to the other, and the value for Area will be given. Start with the formula:
$A_{rectangle} = LW$

One of the variables, L or W, will be in-reference-to the other. If W is in reference to L,
Let Width = (#L +/- #)
where "#" represent given compensation numbers. Substitute the mini-equation for width into the formula and the **template equation** will be like:
$A_{rectangle} = L(\#L +/- \#) = \#$

Or, if W is the variable and L is in-reference-to W, the **template equation** will be like:
$W(\#W +/- \#) = \#$

After the distribution step, this will differ from a perimeter problem because the equation will reveal to be a quadratic equation. Proceed by either factoring & solving or using the quadratic formula. As a quadratic equation can result in two solutions, be sure to weed out the illogical solution. You can throw out a negative value solution because there can't be a negative length or width. You can also throw out a solution whose value is larger than the area because that would be impossible. Finally, plug the value of the solved variable into the mini-equation of the other unknown to find its value. Back to WP37 (pg 177).

Area of a Triangle

Problems involving area of a triangle will give the value of the area and will require you to find the base and height (altitude), both of which will be unknown, and one will be in-reference-to the other. You will use one of the following formulas, which, in essence are the same, they just use a different variable for *height* or *altitude*, which are synonymous:

$$A_{triangle} = \frac{1}{2}bh$$

or

$$A_{triangle} = \frac{1}{2}ab$$

where "h" represents height, "a" represents altitude, and "b" represents the base in either case.

Name your unknowns; "#" is used here to represent the compensation factor. When *base is in-reference-to height*, the **template equation** will be:

$$\frac{1}{2}(h \pm \#)h = A_{triangle}$$

When *height is in-reference-to base*, the **template equation** will be:

$$\frac{1}{2}b(b \pm \#) = A_{triangle}$$

Upon distribution, a quadratic equation will be revealed. Proceed either by factoring & solving or the quadratic formula. Throw out the illogical solution (such as the negative or value larger than the area), and plug the value of the solved variable into the mini-equation of the other unknown to solve for it. Back to WP38 (pg 179).

Areas of Two Squares

Problems like this are built from the formula for area of a square:
$A_{square} = side^2$

I call setups like the following "setting an equation equal to an equation," because you must set

$Area_{square\ 1} = Area_{square\ 2}$ +/- Area compensation number

but you substitute "$side^2$" in for each coinciding Area as:
$(side_1)^2 = (side_2)^2$ +/- Area compensation #

The side of each square will be unknown, so
Let x = the length of the side of one square, and
Let (#x + #) = the length of the side of the other square
where "#" represents compensation factors for *length of a side*, as there
may be a separate compensation factor for *area* as well, as in:
$(side_1)^2 = (\#side_2$ +/- length compensation #$)^2$ +/- Area compensation #

Substitute the unknowns and given compensation values into the
template equation:
$x^2 = (\#x$ +/- #$)^2$ +/- #

Proceed by squaring the binomial on the right side, then combine all
terms (and like-terms) on the same side, set equal to zero, setting up a
quadratic equation in standard form. Continue either by factoring &
solving or by using the quadratic formula. Once complete, throw out the
illogical solution and plug the correct solution value into the mini-
equation of the other unknown to find its value. Back to WP39 (pg 181).

Areas of Two Circles

Problems involving the areas of two circles are very similar to problems involving the comparative areas of two squares, except that the formula for the area of each shape is different. This type is set up by setting an equation equal to an equation as well. The equation will be built from:
$A_{circle} = \pi r^2$

and

$A_{circle\ 1} = A_{circle\ 2}$ +/- Area compensation #

where πr^2 is substituted in for one of the Areas as:

$\pi r^2_{circle\ 1} = A_{circle\ 2}$ +/− Area compensation #

If there is only one unknown such as the radius, r, of one circle, then you can substitute
πr^2 in for the Area of that circle.

Substitute the values of the given radius and the compensation number into the **template equation**:
$\pi(\#)^2 = \pi r^2 \pm \#\pi$

Although you could substitute 3.14 in for pi, it is easier not to because then you can treat pi as a variable, using it to combine like-terms, and then to divide both sides by pi to isolate r^2. Then, r can be solved for by taking the square root of both sides. Accept the positive solution and throw out the negative number. Back to WP41 (pg 185).

Perimeter of a Right Triangle & The Pythagorean Theorem

Anytime a right triangle is mentioned, that is a good indication that the Pythagorean Theorem will be used. Problems like this may seem complicated at first because to build the equation, the Pythagorean Theorem:

$$a^2 + b^2 = c^2$$

must be combined with parts of the Perimeter of a triangle:

$$P_{triangle} = a + b + c$$

We must carefully synchronize all variables so:
a is the length of side a (a.k.a. leg 1),
b is the length of side b (a.k.a. leg 2), and
c is the length of side c (a.k.a. the hypotenuse)

It will be likely that the value of the perimeter will be given and the lengths of all three sides of the triangle will be unknown, but one will be in-reference-to the other.
Let x = the length of "a"
Let (x +/- compensation #) = the length of "b"
Let c remain "c"

Before substituting the unknowns with x in, rearrange the formula of Perimeter, subtracting a and b from both sides to solve for c as:

$$c = P - a - b$$

Next, substitute "P − a − b" in for c in the Pythagorean Theorem, and keep in mind that c is squared in the Theorem, so the quantity substituted in for c will also be squared:

$$a^2 + b^2 = (P - a - b)^2$$

Now substitute the unknowns for sides a and b in, as well as the given value for P into the **template equation**:

$$x^2 + (x +/- \#)^2 = [\# - x - (x +/- \#)]^2$$

Simplify carefully. Notice there is a binomial to be squared on the left side. On the right side, in the brackets, simplify and combine like-terms, which will also turn into a binomial to be squared. Move all terms to one side to make a standard form quadratic equation, then attempt to factor & solve or use the quadratic formula. Throw out the illogical solution for x, then substitute the correct solution into the mini-equation to solve for the length of "leg 2." Then, take the values of a and b and substitute them into the Pythagorean Theorem to solve for c. Back to WP40 (pg 183).

Similar Right Triangles, Cast Shadow, Using Proportion/LCD

When you see a shadow problem, you should automatically think *proportion of two right similar triangles*. Name the unknowns:
Let x = the length of one side of one triangle
Let (x +/- compensation factor) = the length of a coinciding side

Set up a proportion using two consistent, similar sides of the triangles, for instance:
$$\frac{Base_{larger}}{Height_{larger}} = \frac{Base_{smaller}}{Height_{smaller}}$$

but the hypotenuse may be used instead of one of the legs. Then substitute the unknowns for two sides and the given values for the other two sides in to the **template equation**:
$$\frac{\#}{x \pm \#} = \frac{\#}{x}$$

Cross multiply or multiply both sides by the LCD. After simplifying and solving, plug the value of the x into the mini-equation to find the value of the other unknown. Back to WP42 (pg 187).

The Area or Volume of the Shaded Region

There is not much to say about these except to use the formulas of the given shapes, and to *subtract the inner shape from the outer shape*. If there are multiple inner shapes, subtract the areas or volumes of all the inner shapes.

For an unconventional shape made from smaller connected shapes, use the equations of areas or volumes of each shape and *add* their areas or volumes together.

There's also a chance you might have to add and subtract the areas or volumes of various shapes together. Try to break these types of problems down in pieces.
Back to WP43: Area of the Shaded Region (pg 189).
Back to WP44: Volume of the Shaded Region (pg 191).

RATE OF SPEED

Rate, generally, is a ratio of some unit measurement per unit time. One specific type of rate you often encounter in word problems is *rate of speed*. Some common units of rate of speed are

$$\frac{\text{miles}}{\text{hour}}, \frac{\text{kilometers}}{\text{hour}}, \text{and } \frac{\text{meters}}{\text{second}},$$

and there are many others.

There are a number of types of word problems you will encounter that are based on the simple *rate of speed* formula:

$$\text{rate} = \frac{\text{distance}}{\text{time}}$$

or, abbreviated as:

$$r = \frac{d}{t}$$

This equation can be rearranged based on what values are given and what is unknown. If the distance is unknown, the equation can be rearranged by cross-multiplying to get:

$$d = rt$$

From here, if time is the unknown, divide both sides by r to get:

$$t = \frac{d}{r}$$

Earlier on, you may encounter simple problems in which the values for two of the variables (r, t, or d) are given and you must solve for the other. When doing problems with simple formulas, you can either plug the given values into the formula, then rearrange it to solve for the unknown variable, or you can rearrange the formula to solve for the unknown variable first, then plug in given values and simplify.

In more complicated problems, however, such as problems involving more than one unknown, this may not be as pragmatic because compensation factors are used to describe one unknown in-reference-to the variable, and these compensation factors can be different from problem to problem. For this reason, algebraic techniques must be utilized to simplify the equation to solve for the unknown variable after the givens are plugged-in. Also for this reason, there are no set formulas, however, the template equations are built from the simple rate of speed formula.

The following *compensation words* are used in rate of speed related problems:

- If an unknown is a certain amount **slower than** the reference variable, *subtract* that amount *from* the reference variable.

- If an unknown is a certain amount **faster than** the reference variable, *add* that amount *to* the reference variable.

- If an unknown is a **multiple faster than** the reference variable, multiply that multiple times the variable. For instance, if x = speed of car 1, and car 2 is going double the speed of car 1, let car 2 = 2x.

Again, as you begin to encounter more complicated problems, you will have to use variations of the rate of speed formula. This book will provide you with templates for most of the types you will encounter, but maybe not every type. The logic explained in this section will help you set up your equation for the types not covered in this book. Let's proceed with the types that are…

Two Trains Leave the Station, When Will They Meet?
Time is Unknown, Total Distance Given, Different Speeds

In problems like this where two vehicles are traveling towards each other and time is unknown but the same for each traveler, pay close attention to the total amount or amounts given. Usually, if a *total* amount is given, it is likely the sum of two quantities. In this type of problem, total distance is given, so we can start building the equation as:
$d_1 + d_2 = d_{total}$

where two distances are added to equal the total distance. Since each "d" is different, we've attached subscripts to differentiate one from the other. In this type of problem, time is unknown, and even though there are two people or vehicles, the times for both will be the same. In order to get "t" into the equation, we will use the rearrangement of the rate of speed formula, solved for d:
$d = rt$

and substitute "rt" in for each "d":
$r_1t + r_2t = d_{total}$

Since each "d" is different, but each "t" is the same, we've attached subscripts to each "r" to distinguish one from the other, as well as to associate each "r" with its coinciding "d". This makes the **template equation**:
#t + #t = #

Where the "#"s represent the given values of each rate of speed and the total distance. Combine like-terms to get
#t = #"
then divide both sides by the coefficient in front of "t" to find its value. If the question asks for a specific clock-time, you may need to convert the decimal part of hours to minutes, then put into clock form: hh:mm.
Back to WP23 (pg 143).

Catch-Up: Traveling In The Same Direction, When Will They Meet?

In problems like this where the travelers depart from the same starting point and meet (or pass) at the same point, it is implied that the distance each person traveled is the same, however, the value of the distance probably will not be given because it isn't needed. But the travelers will have left at different times, and you will be told the time-difference.

Here's where the equations come from:
Since the *distances are implied to be the same*, but not given, rearrange the rate of speed formula for d: $d = rt$

Now put subscripts in to differentiate between each person:
$r_1t_1 = d = r_2t_2$

Since they both equal the same d, take out the d and just set the two "rt"s equal to each other:
$r_1t_1 = r_2t_2$

From here, two equations can be built (but you only need one), all depending on how the unknowns are named, specifically, depending on which person or vehicle you assign as the reference variable.

If it is said that the person who left first left a certain number of hours *before* or *ahead* of the other, this infers subtraction from the time of the person who left last, so,
Let t = the time of the person who left last. Then the other unknown, by default, would be:
Let (t – the time difference) = the time of the first person to leave

Or, if it is said that the person who left last left a certain number of hours *behind or after* the other, this infers addition to the time of the person who left first. In this case, you will have to
Let t = the time of the person who left first so you can add to the time of the first persons as:
Let (t + time difference) = the time of the last person to leave

This makes logical sense because the person who leaves first will be driving for a longer time at a slower speed, and the one who leaves last and catches up to the other will drive for a shorter time at a faster speed.
Continued on the next page...

Plug the assigned unknowns in for each respective "t", into:
$r_1 t_1 = r_2 t_2$

depending on how you named them. The **template equations** will be different, coinciding with whether the *before* or *after* reference was used.
For the *before*, use: $\#(t - \#) = \#t$
For the *after*, use: $\#t = \#(t + \#)$

For either equation, proceed by distributing through the parentheses, combining like-terms, and solving for t. Then, plug the value of t into the mini-equation of the other unknown to find its value.

Back to WP24 (pg 145).

Two Different Roads: Rates are Unknown, Total *Distance* Given

Two Different *Roads* can also mean different *conditions*, such as different
- terrains,
- gas mileage,
- pathways, or even
- different vehicles

Problems like this are built from:
distance$_1$ + distance$_2$ = distance$_{total}$

and the rate of speed formula solved for d: d = rt

where "rt" is substituted in for each "d" as:
$(t_1)(r_1) + (t_2)(r_2) = d_{total}$

Let x = the rate of speed on one road, and
Let (x +/- compensation factor) = the rate of speed on the other road

Substitute the given values for total distance and each "t", and the unknowns for each respective "r" into the **template equation**:
#x + #(x +/- #) = #

Note: The given values for unknowns r and t are written in the order "(t)(r)" in the template equation because the value of t is given, and we usually write numeric coefficients first followed by their associated variable or group factors.

Proceed by distributing, combining like-terms and solving for x. Then, plug the value for x into the mini-equation of the other unknown to solve for its value. Back to WP25 (pg 147).

Two Different Roads: Rates are Unknown, Total *Time* Given

Two Different *Roads* can also mean different *conditions*, such as different
- terrains,
- gas mileage,
- pathways, or
- different vehicles

This equation is built from:
time + time = time

or more specifically:
$t_1 + t_2 = t_{Total}$

and the rate of speed formula solved for t:
$$t = \frac{d}{r}$$
where $\frac{d}{r}$ is substituted in for each t as:

$$\left(\frac{d}{r}\right)_1 + \left(\frac{d}{r}\right)_2 = t_{Total}$$

Let x = the rate of speed travelling on one road
Let (x +/- compensation #) = the rate of speed travelling on the other road

Substitute the unknowns in for each rate:
$$\frac{d_{one\ road}}{x} + \frac{d_{the\ other\ road}}{x \pm \#} = \text{total time travelled}$$

Plug in the given values for each distance and the total time travelled into the **template equation**:

$$\frac{\#}{x} + \frac{\#}{x \pm \#} = \#$$

Be sure to do any necessary conversions for units of time so they are the same (they will likely all need to be in *hours*), either in the beginning or end of the problem. Back to WP26 (pg 149).

75

Rate of Speed Unknown but Same; Distances Given, Time Not Given but Reference to Time Given

This equation is built from:
time = time

or more specifically:
$time_{route\ 1} = time_{route\ 2} +/-$ time compensation #

and the rate of speed formula solved for t:
$$t = \frac{d}{r}$$

where $\frac{d}{r}$ is substituted in for each t as:

$$\frac{distance_1}{rate} = \frac{distance_2}{rate} \pm time\ compensation\frac{\#}{\#}$$

Because the rates are the same, there will only be one unknown:
Let x = the average rate of speed of the paperboy

Also, notice that the compensation time number is shown as a fraction. This is because the problem may give this value as a fraction, and if it is, you can use its denominator as part of the LCD towards simplifying. Or, you could convert it to decimal form. Plug the values for distance and the compensation time number, and x for each rate into the **template equation:**

$$\frac{\#}{x} = \frac{\#}{x} - \frac{\#}{\#}$$

Proceed by multiplying each term by the LCD, which will then set up a linear equation. Back to WP27 (pg 154).

Rate of Vehicle Unknown; Distance & Rate of Current Given; Same Time

The vehicles involved in problems like this are usually planes flying through the air or boats on moving water such as a river. The equation used for this type of problem is built from:
time = time

or, more specifically:
$\text{time}_{\text{downstream}} = \text{time}_{\text{upstream}}$

and the rate of speed formula solved for t:
$$t = \frac{d}{r}$$

where $\frac{d}{r}$ is substituted in for each t as:

$$\left(\frac{d}{r}\right)_{\text{downstream}} = \left(\frac{d}{r}\right)_{\text{upstream}}$$

Let x = the rate of the vehicle in no current
Let (x + given rate of the current) = the rate of the vehicle moving downstream
Let (x − given rate of the current) = the rate of the vehicle moving upstream

As the value of the rate of the current will likely be given, plug it into the unknowns. Then, substitute the unknowns and the given values of the distances into the **template equation**:

$$\frac{\#}{x + \#} = \frac{\#}{x - \#}$$

Proceed by finding the extraneous solutions, then multiply both sides by the LCD or cross-multiply. Back to WP28 (pg 156).

Rate of Vehicle & Current Unknown (Upstream/Downstream)

Unlike most of the other problems involving rate of speed, this type of problem will have two variables and will use a system of two linear equations to solve. The variables and unknowns will always be the same for problems of this type, which is why you can use these **template unknowns**:

Let x = rate of speed of the vehicle in no current
Let y = rate of speed of the current
Let (x + y) = rate of speed of vehicle travelling *with* the current (downstream)
Let (x - y) = rate of speed of vehicle travelling *against* the current (upstream)

Each equation will be built from the rate of speed formula, solved for d:
rt = d
Since the values for time are given and the variables are in a parenthetical group, the "rt" will be re-positioned as "(t)(r)" so the numbers given for time can be written as a numeric coefficient to the parenthetical groups. Plug in the values of each respective given time and distance, and the coinciding unknowns into the **template equations**:
1) #(x + y) = #
2) #(x − y) = #

Note: For problems like this, you can default to setting up these same template unknowns, and the equations will be alike as well, except for the different given values for distance and time.

The first step towards solving will be to divide both sides by the coefficient in front of the unknown binomial. (Or, you can distribute the coefficient number into each set of parentheses, which will just result in larger numbers). Proceed in the next step by utilizing either the Substitution Method or the Addition/Elimination Method. Once you find the values of x and y, be sure to plug the values of each back into each mini-equation for the unknowns of the rate of speed of the vehicle moving upstream and downstream, if the question asks for them. Back to WP29 (pg 158).

Travel Times Unknown; References to Rate & Time Given

The equation for problems like this are built from:
$$\text{rate}_{\text{vehicle 1}} = \text{rate}_{\text{vehicle 2}} +/- \text{rate compensation } \#$$

and the rate of speed formula solved for rate:
$$r = \frac{d}{t}$$

where $\frac{d}{t}$ is substituted in for r of each vehicle:

$$\frac{d}{t} = \frac{d}{t \pm \text{time compensation}} \pm \text{rate compensation}$$

Let x = the time for vehicle 1 to travel
Let (x +/- time compensation #) = the time for vehicle 2 to travel

Since there are two compensation numbers given for this type of problem, be sure to put each given value in the proper place. The compensation number for *time* will already be written as part of one of the unknowns (in a denominator). The compensation number for *rate* must be added or subtracted appropriately on one side of the equation. Substitute the given values and unknowns into the **template equation**:

$$\frac{\#}{x} = \frac{\#}{x \pm \#} \pm \#$$

Back to WP30 (pg 161).

Note: Rate can be used as a shell equation for other units such as "miles per gallon," as seen in WP51 (pg 205).

SPLITTING A TASK

In problems like this, "task" or "job," as well as other words, or even specific tasks can be used interchangeably. There are two main template equations used for *splitting a task* problems. One form is built around the *fraction of the task completed per hour* and the other is built around the *time to complete a whole task*. Either form creates rational equations. The typical problems involve two people, however the two equations below can apply to more than two people by adding appropriate fractions to the equation.

The form built from *the time to complete a whole job* is:

$$\frac{\text{Time to complete task together}}{\text{Time for person 1 to do task alone}} + \frac{\text{Time to complete task together}}{\text{Time for person 2 to do task alone}} = 1$$

In words: (The time it takes to split the task working together *over* time for person 1 to do the whole task alone)
plus (the time to split the task working together *over* the time for person 2 to do the whole task alone)
equals 1 (in reference to one whole task)

This form of the equation is the recommended one and is used in the annotated examples in this book because it is easier to use. Notice that, in this form of the equation, the sum of the fractions *always equals 1*. Also notice that the *numerator of both fractions is the same*: the time to complete the task together. If that is the unknown, then both numerators will be x, as seen in the **template equation for when the time to complete the task *together* is unknown:**

$$\frac{x}{\#} + \frac{x}{\#} = 1$$

However if the *time to complete the task when working together* is given, put that number in *both numerators* and put the variables and/or unknowns into the appropriate places in the denominator(s), as shown in the next two scenarios.

Continued on the next page...

If the **time for one person to complete the task *alone* is unknown,** use this **template equation:**

$$\frac{\#}{\#} + \frac{\#}{x} = 1$$

If the **time for *each* person to complete the task alone is unknown,** use this
template equation:

$$\frac{\#}{x} + \frac{\#}{x \pm \#} = 1$$

Proceed by multiplying each term by the LCD to get rid of the fractions. This is a 1st degree equation. Combine the x-terms on one side and the constants on the other, divide both sides by the coefficient in front of x and solve for x. Then, plug the value for x into the other unknown to find its value.

If the **fraction of the task completed in 1 hour is unknown,** whether by working together or for each person, take the reciprocal of the time to complete the whole task. For instance, if you find the time to complete the whole task working together is
$\frac{5 \text{ hours}}{1 \text{ task}}$, flip it to $\frac{1 \text{ task}}{5 \text{ hours}}$, then

divide 1 by 5 to get $\frac{\frac{1}{5} \text{ or } 0.2 \text{ task}}{1 \text{ hour}}$.

The **other form equation** is built from the *fraction of the task completed per hour*:

In words: (Fraction of the job person 1 can do in one hour) plus (Fraction of the job person 2 can do in one hour) equals (Total fraction of the job that can be done in one hour when working together)

The **template equations** in this form will look like:

$$\frac{\#}{\#} + \frac{\#}{\#} = \frac{\#}{\#}$$

with givens, the variable, and unknowns in the appropriate place(s). Notice that, in this form, it equals a fraction (not 1 as in the other form). In this form, all the fractions are alike by having the common denominator unit of "1 hour".

Which form of equation should you use? The *time to complete a whole job* form is recommended, however you might choose based on the information given and what the question asks for in the problem. If "fraction of task per hour" is given, use the "fraction of task per hour" template equation. If "times to complete a whole task" are given or asked for, use the "times to complete a whole task" template equation."

What if the forms are mixed in the problem? You can use either equation, then give the reciprocal of the final answer. For instance, if you find "fraction of the task completed in 1 hour when working together," the reciprocal of that will be "time to complete the whole task when working together," and vice versa. If you do need to take the reciprocal, you will need to perform the division of the numbers in the fraction. Suppose, for example, the question asks to find the *time it takes to complete a task when working together*, but you solve to find the *fraction of the task completed in one hour when working together* is:

$$\frac{\frac{1}{4} \text{ task}}{1 \text{ hour}} \text{ or } \frac{0.25 \text{ task}}{1 \text{ hour}}. \text{ Flip it to } \frac{1 \text{ hour}}{0.25 \text{ task}}, \text{ then}$$

divide 1 by 0.25 to get $\frac{4 \text{ hours}}{1 \text{ task}}$.

Back to WP31: Time to Complete Task Working Together (pg 164).
Back to WP32: Time to Complete Task Alone Unknown (pg 166).
Back to WP33: Time for Each to Complete Task Alone Unknown (pg 168).

82

MIXED ITEMS: Nuts, Candy, Coins, Tickets, Chemicals, Investments

With the exception of mixes involving chemicals or solutions, many *mixed items* problems are built around the dollar value of each item, where

$$\text{dollar value}_{item\ 1} + \text{dollar value}_{item\ 2} = \text{total dollar value}$$

and

$$\frac{\text{dollar value}}{\text{item}}(\# \text{ of items}) = \text{dollar value}$$

Since, usually, the value of *each* item and the total *combined* value are given, and the numbers of each of the two items are unknown, substitute

$$\frac{\text{"dollar value}}{\text{item}}(\# \text{ of items})\text{" in for "dollar value"}$$

into the first equation from above, becoming:

$$\left(\frac{\$ \text{ value}}{\text{item}}\right)_1 (\# \text{ of items})_1 + \left(\frac{\$ \text{ value}}{\text{item}}\right)_2 (\# \text{ of items})_2 = \text{total } \$$$

In the mini-equation above, the units "item" cancels with "items". Since, usually, the number of each of the two items are the unknowns, set up the unknowns accordingly:

Let x = the number of items of item 1
Let (x +/- compensation #) = the number of items of item 2
or, if *total number of items* are given:
Let (total # - x) = the number of items of item 2

Plug the given dollar value in times each associated unknown. When a *compensation factor* relates one unknown in-reference-to the other, the **template equation** will be:
#.##x + #.##(x +/- #) = #.##

When the *total number of items* is given, the **template equation** will be:
#.##x + #.##(# - x) = #.##

For mixtures involving chemicals, percentages converted to decimal form will be used instead of dollar value, as seen in WP9 (pg 121).

These types of problems are linear equations which involve:
- distribution,
- combining like terms, which will likely involve adding two decimal form numbers together,
- then solving for x.

Once you find x, you must substitute that value into the other unknown *mini-equation* to find the "remaining amount."

Two Coins of Different Value (One Variable)

The equation you will use is built from:
dollar value$_{coin 1}$ + dollar value$_{coin 2}$ = total dollar value of all coins

where
$$"\frac{\text{dollar value}}{1 \text{ coin}} (\# \text{ of coins})"$$

is substituted in for the "dollar value" of each coin from the equation above.

for as many coin denominations there are. This equation applies to two coin denominations:
(value of coin 1)(# of coins 1) + (value of coin 2)(# of coins 2) = value of all coins

Set up the unknowns accordingly:
Let x (or the first letter of the name of the coin) = the number of coins 1
Let (x +/- compensation #) = the number of coins 2
or, if total number of coins are given:
Let (total # - x) = the number of coins 2

Substitute the appropriate unknowns, each coins respective dollar value, any given compensation numbers, and the given total value of all coins into the **template equation**:
0.##(x) + 0.##(#x +/- #) = #.##

The value of each coin should be expressed in dollars (meaning in decimal form). If only the total value of the coins is given, a compensation number will also be given to show one unknown in-reference-to the other unknown (variable) and can be solved using the template equation above.

If two totals are given (the total *value* of the coins and the total *number* of coins), the other unknown (reference-to-the variable) will be "(total coins − x)" and can be solved with the **template equation**:
0.##(x) + 0.##(# - x) = #.##

Back to WP5 (pg 113). To see a two-coin problem solved with two variables and two equations, see WP45 (pg 193).

To see a three-coin problem with one variable, a reference-to-the variable, and a reference-to-a-reference-to-the variable, see WP6 (pg 115).

To see a three-coin problem using three variables and three unknowns, see WP52 (pg 209).

Two Different Priced Tickets (One Variable)

Problems involving two different priced tickets are another form of problem related to a mix of two items of different value. The equation is built from:

dollars$_{\text{ticket 1}}$ + dollars$_{\text{ticket 2}}$ = dollar value of combined tickets

Let x = the number of one type of ticket at its given price
Let (total − x) = the number of the other type of ticket at its given price

Multiply the price of each type of ticket times its unknown to make the more specific equation:

(ticket price$_1$)(# of tickets$_1$) + (ticket price$_2$)(# of tickets$_2$) = total revenue

Substitute the given prices attached to their unknowns, the given total number of tickets, and the total revenue into the **template equation**:
#x + #(# − x) = #

Solve for x then substitute its number in for x into the mini-equation of the "number of the other type of tickets."

Back to WP7 (pg 117). To see a two-ticket problem solved with two variables and two equations, see WP46 (pg 195).

Buying Mixed Items at Two Different Unit Prices (One Variable)

Problems involving the purchase of mixed items at two different unit prices are built from the base equation:

dollars$_{item\ 1}$ + dollars$_{item\ 2}$ = total dollars spent on combined items

where unit price is given as $\dfrac{\$}{lb}$ and the amount of money *spent* is found by multiplying the unit price by the weight in pounds:

$$\left(\frac{\$}{lb}\right) lb = \$ \text{ spent. Then,}$$

"$\left(\dfrac{\$}{lb}\right) lb$" can be substituted in for dollars

into the first equation above, as:

$$\left(\frac{\$}{lb}\right)_1 (lb)_1 + \left(\frac{\$}{lb}\right)_2 (lb)_2 = \$$$

Let x = the weight of one type of item at its given price
Let (total weight – x) = the weight of the other type of item at its given price

Substituting the unknowns in makes the more specific equation:
(unit price)$_1$x + (unit price)$_2$(total weight – x) = total dollars spent

Substitute the given unit prices attached to their associated unknowns, and the given value for total amount paid on combined items into the **template equation**:
#.##x + #.##(# - x) = #.##

Back to WP8 (pg 119). To see a similar problem solved with two variables and two equations, see WP50 (pg 203).

Mixing Two Chemicals to Make a Final Solution (One Variable)

Problems involving mixing chemicals will be set up slightly differently than the former three problem types, for reasons explained below. From a unit-perspective, the equation is built from:

$\text{Liters}_{\text{chemical 1}} + \text{Liters}_{\text{chemical 2}} = \text{Liters}_{\text{final mixture}}$ (see Note below)

Let x = the volume of a solution at a given percentage
Let (total volume – x) = the volume of the other solution of the other given percentage

In this type of problem, there is one intermediate mini-calculation that must be done to complete the setup that doesn't need to be done in other *mix* problems. As the total volume is usually given, and the percentage of the final mixture is given, you must multiply the two numbers together as seen on the right side of the equation below:

$(\% \text{ in dec. form})(\text{vol.})_1 + (\% \text{ in dec. form})(\text{vol.})_2 = (\% \text{ in dec. form})(\text{total vol.})_{\text{final}}$

which will yield a *new number* (see right side of the equation below):

0.##x + 0.##(total volume – x) = product of final solution

Plug in the percentages in decimal form attached to their unknowns, and the given total volume and the product of the final solution explained above, into the **template equation**:

0.##x + 0.##(# – x) = #.##

Note: Usually, the units for the values in the equation are clear. For this type, you could say they are "Liters of the active ingredient," however, they are not so obvious in this particular type of problem. In short, this is because *chemical concentration* can be expressed in many different units, including percentage. Even *concentration by percentage* can be expressed with respect to different units such as weight, volume, and moles (atoms or particles). For simplicity, it's best to just use the template equation, plug in the given values, and multiply the percents in decimal form times each respective volume (which will initially be the unknowns); just make sure each volume is in the same unit as each other (likely liters or milliliters).

Back to WP9 (pg 121). To see a two-chemical mix problem solved with two variables and two equations, see WP48 (pg 199).

Investing in Two Simple Interest Investments (One Variable)

Investing in Two Different Simple Interest Investments problems are really *mixed items problems* involving two unknowns and one variable (we just don't think of investments as "items") and are much more like the chemical mixture problems because the percentage for yearly interest rate is inserted in the equation in decimal form. The unknowns are modeled the same as any from the *mixed items* type.

Let x = the amount of money invested at a given percentage
Let (total amount invested – x) = the amount of money invested at the other percentage

The equation is built from:
$\text{Interest}_{\text{Investment 1}} + \text{Interest}_{\text{Investment 2}} = $ total Interest earned

and the formula for simple interest:
I = Prt

where "Prt" is substituted in for each respective "I" in the first equation above. Since "t" is assumed to be "1", you can think of "Pr(1)" or just "Pr" as being substituted in for each I:
$(Pr)_1 + (Pr)_2 = $ total interest earned

Additionally, since the rates are given and each "P" is unknown, the factors are written as "rP": $(rP)_{\text{Investement 1}} + (rP)_{\text{Investment 2}} = $ total Interest earned

so the rate as a percentage in decimal form is written as the coefficient and attached to the left of the variable and unknown factors:
(% in dec. form)x + (% in dec. form)(total amount *invested* – x) = total amount *earned*

Plug the unknowns, the percents in decimal form, the given total amount invested, and the total amount earned into the **template equation**:
0.##x + 0.##(# – x) = #

Back to WP22 (pg 141). To see a simple multiple-investment problem like this solved with two variables and two equations, see WP49 (pg 201).

TWO VARIABLES, TWO EQUATIONS

When you are doing Systems of Two Linear Equations in class, you are also likely to be given word problems which will use systems of two linear equations to solve. As you would expect, the problem is dependent on properly naming the variables and setting up *two equations*. Always look for *two total amounts* given in the problem, as these will likely be what each of the two equations will be set equal to (see Note 1 below). Also, look for each total to be in different units (perhaps one in *dollars* and the other in *number of items* or *pounds*). The units of the given totals will give you insight on the units to build each of the two equations around.

As you may remember, the "solution" to a system of two linear equations is the point at which two lines intersect on a graph, also known as an ordered pair in the form (x, y), which is why the solutions are given this way in each example. For word problems, the solution, or graphical point of intersection, is also known as the *equilibrium point*.

The **template equations** may look like:
1) $x + y = \#$
2) $\#x + \#y = \#$

or

1) $\#x + \#y = \#$
2) $\#x + \#y = \#$
with different numbers,

There's also a chance one of the equations may only contain one variable, such as:
1) $x = \#$
2) $\#x + \#y = \#$
which is a good indication to use the substitution method, or

there's a chance the variables can be in the denominator or numerator, as:

$$\frac{\#}{x} + \frac{\#}{y} = \#$$

or

$$\frac{x}{\#} + \frac{y}{\#} = \#$$

as in WP51 involving *Miles per Gallon* (pg 205 & pg 93).

Once the two equations are set up, you can use either the Substitution Method or Addition/Elimination Method to solve.

Most examples in this section are an alternative method to the same problem solved with one variable and one equation (in some cases, a piece of information is added and/or removed). In those examples, the equation used is *one* of the equations used in the system of two linear equations version, but the unknowns will be slightly different. You can compare the answers of each companion problem solved by both methods. All additional information needed for each *two variables, two equations* problems already accompany each example.

Note 1: There is one exception to this which is WP48: Mixing Chemicals (pg 199) in which you must *calculate* the *final volume* (as shown in its one-variable companion problem WP9, pg 121).

Below are links back to each problem:
Back to WP29: Upstream/Downstream: Rate of Vehicle & Current Unknown (pg 158).
Back to WP45: Two Coins of Different Value (2 Variables), pg 193.
Back to WP46: Two Different Priced Tickets (2 Variables), pg 195.
Back to WP47: Manufacturing Two Different Items (2 Variables), pg 197.
Back to WP48: Mixing Two Chemicals to Make a Final Solution (2 Variables), pg 199.
Back to WP49: Investing in Two Simple Investments (2 Variables), pg 201.
Back to WP50: Buying Mixed Items at Two Different Unit Prices (2 Variables), pg 203.

Note 2: Solving Systems of Two Linear Equations is covered more extensively in *ALGEBRA IN WORDS*, and solving Systems of Two Linear Inequalities are covered in *ALGEBRA IN WORDS 2*.

See Also: WP52 involving 3 variables and 3 equations (pg 209 & pg 94).

Miles per Gallon (2 Variables; 2 Setups)

Problems like this are set up in much the same way as rate of speed problems, however, may seem a bit more confusing for two reasons:
- they can be set up in two different ways, and
- the equations for either way can be a bit unit-intensive

The two ways they can be set up are either where the variable represents miles or gallons, and the equations set up are dependent on which units you let the variables represent at first. The equations will be different, and one method (shown in the Alternative "gallons" Setup in WP51) requires two conversions near the end of the solving process.

The best advice is to just follow the template equations once you recognize the type of problem, however, it's still important that you know where the equations come from. All other equations are explained in good enough detail in WP51, except for equation 2 of the "miles" setup. Equation 2 is built around:

$gallons_{highway} + gallons_{city} = $ total gallons

and:

$$\frac{miles}{gallon} = \text{gas rate}$$

by solving (rearranging) for gallons by cross multiplying, then dividing both sides by "gas rate" to get:

$$\frac{miles}{\text{gas rate}} = \text{gallons.}$$ Substitute $"\frac{miles}{\text{gas rate}}"$

in for each "gallons" in the equation above to make:

$$\frac{miles_{highway}}{\text{gas rate}_{highway}} + \frac{miles_{city}}{\text{gas rate}_{city}} = \text{total gallons}$$

Since the variables in this setup represent miles and gas rates and total gallons are given, the **template equation** for equation 2 is:

$$\frac{x}{\#} + \frac{y}{\#} = \#$$

Back to WP51: Distances on Two Different Roads Unknown, Miles/Gallon Given (2 Variables), pg 205.

THREE VARIABLES, THREE EQUATIONS

You will need to set up three equations to solve a system of three linear equations with three variables. Use the same logic in problems with two variables and two equations to set up the equations. However, three totals may not be given. That's okay because not all three equations have to have all three variables in them. The problem may give clues comparing one variable in-reference-to another with compensation factors. Use the procedure given below:

Procedure for Solving a System of 3 Linear Equations with 3 Variables

Solving a system of three linear equations with three variables is done by using the same techniques as for solving a system of two linear equations with two variables, in multiple steps. This procedure is written assuming all three variables are used in each of the three equations (see Note below).

1. Name each of the three variables.

2. Use the given totals and compensation factors to set up three equations using the named variables. Number each equation as 1, 2 & 3 (again, see Note below).

3. In each of the next two steps, two of the three equations are paired up and the Addition/Elimination method is applied in order to (eliminate the same variable in each pair of equations to) create *two new equations* with the *same two variables*.

3a. Perform the Addition/Elimination method to two equations, such as 1 & 2, causing terms of the same variable to cancel out (say, the x-terms). This creates what we'll call (new) equation 4, in terms of y & z.

3b. Perform the Addition/Elimination method to another pair of equations, such as 1 & 3 (or, could be to 2 & 3), causing the terms with the same variables to cancel out as in Step 3a, (again, say, the x-terms). This creates (new) equation 5 also in terms of y & z.

This now creates a *system of two linear equations* with (the same) two variables. However, if the same variable was not cancelled out in Steps 3a & 3b, the next step will not work. Continued on the next page...

4. Solve this set of two new linear equations (4 & 5) using either the Addition/Elimination Method or the Substitution Method[*], yielding the values of two variables (in this hypothetical example: y & z).

5. Plug both of these values back into one of the original three equations to find the value of the third variable (in this hypothetical example, plug the values of y & z into equation 1 to solve for x).

Note: When All 3 Equations Do Not Contain All 3 Variables:
Each of the three equations may not have all three variables. (You can detect this because one instruction may describe one variable in-reference-to another using compensation factors.) If this is the case, this is fine because it will *save you a step* (Step 3b... you will not have to do the Addition/Elimination method to two pairs of equations... you will only have to do it to one pair, the pair with all three variables, making only "new equation 4," while making sure to *eliminate the variable which is also missing from the original equation* unused in Step 3a). In Step 4 treat new equation 4 and the original (so far unused) equation with the same missing variable as a *system of two linear equations with two variables*, and continue to Step 5.

For example, let's say original equation 3 only has terms with x & y (z is missing). Do Step 3a to equations 1 & 2 causing the z-terms to cancel out, creating equation 4. Perform Step 3a to equations 4 & 3, solving for the x & y values. Then, in Step 5, plug in the values for x & y into equation 1 (or 2) to find the value of z. This type of scenario is the case in WP52 (pg 209).

*The Substitution and Addition/Elimination methods are explained in more depth with detailed examples in *ALGEBRA IN WORDS*.

EXPONENTIAL FUNCTIONS
Exponential (Continuous) Growth & Decay (Half-Life)

Exponential growth is also known as *geometric growth*.
The formula for exponential *growth* and *decay* are the same:
$f(t) = A_0 e^{kt}$ or $A = A_0 e^{kt}$

It is important to know what each symbol represents.
A = The Amount at a certain time. This amount varies as the other
variable, t, varies, which makes A a function of time. Therefore, f(t) can
be replaced with A, and vice versa.

A_0 = The Initial Amount. This number is a constant. It is worth noting
that the attached subscript "0" represents "at time = 0" meaning "before
the process begins." This subscript, as any subscript, is not a number
used in the equation or calculation, it is only a descriptor to A.

A_f = The Final Amount, as used in Growth problems, seen in Step 2a.

e = Euler's Number, often called "the number e." This number is also a
constant, 2.72, when rounded to the nearest hundredths place, however,
the actual number does not need to be used in growth and decay
problems because taking the natural log of e cancels it out (as seen in
Steps 1f and 2e).

k = The growth or decay rate constant. This is a constant (number)
relative to a defined time period for each specific type of problem. If the
sign of k is positive, it is a growth problem. If the sign of k is negative, it
is a decay problem (when k is negative, think of "e^{kt}" in the denominator
under "1" with a now positive k). Unless k is given, it is often meant to
be solved for in Step 1 of the problem. Note: Sometimes "r" is used in
place of k, as seen when the Continuous Growth model is used for
Compounding Interest. You may not know the sign of k to start
(although, based on the questions asked, you might be able to make a
correct assumption). Regardless, to do the problem, it actually doesn't
matter, as you will probably solve for k anyway, and the sign will reveal
itself then.

t = time, but it's important it is recognized in the proper context. Time in
this context should be thought of as *time elapsed,* or *duration.* Time, as t,
can vary from zero to infinity, and is associated with the changing values
of "A" (since A is a function of t). "Time" is not the same as "the year,"

even though time is often in *units of years*. The starting, given *year* must be calibrated for "t" because the start time is always "t = 0". For instance, if the start year is 1980, t = 0 at 1980, and likewise, t = 1 for 1981, etc. In Growth and Decay problems, when the value of A is asked for at a certain year, you don't put the year in for t, you put the time elapsed, or years since the start year in for "t". You can always use this simple equation to find the correct value of t:

$t = Year_{asked} - Year_{initial}$

Likewise, if you are asked in what year A will reach a certain given value, you first must solve for t, then add that number to the starting year, as seen in Step 2g of the procedure.

It is worth noting that at $t = 0$, $A = A_0$.

A typical problem will likely have multiple questions which will include:
1. What is the exponential function (for given conditions)?
2. By what year will "A" reach a certain given number?
There may also be a third question, such as: What will A be at the end of a given year?

Overview of the Steps
In order the answer the questions above, word problems and questions involving this formula are solved in about 2 parts:
1. Solving for k to find the function (sometimes called "model") of a certain time period, which will be used for the next question.
2. Using the function from Part 1 to solve for "t", given a certain value (A) in reference to the initial Amount, A_0.

More Valuable Information
Since problems like this often deal with large numbers, such as populations, the numbers given are often given in *bundle-words*, such as # thousand or # million, and a question might ask: How many *millions*…? It is important to only put the number-portion into the formula, not including the bundle-word. In other words, if a problem says that the initial amount (A_0) is "76 million" and asks "How many millions…?" (as in WP53, pg 212), you put "76" (*not* "76,000,000") in for A_0, because the model is already in the context of millions. Likewise, if graphed, the (y) axis for population will be scaled by the prefix numbers, where each hash-mark may represent a million, or even fives or tens of millions. In this case, the units of the axis will be defined as "population in millions."

Procedure for Exponential Growth & Decay

Part 1: Find k and give the function with the value of k

Into the original formula,

1a. Plug the given number representing the *future amount* in for A (on the left).

1b. Plug the number for *Original Amount* in for A_0.

1c. Find and plug in the value for "t". Use the formula:
$$Year_{asked} - Year_{starting} = t$$

Note: Although you might feel compelled to substitute 2.72 in for "e", you should leave e as e. Since you will be solving for "k" (a variable in the exponent), you will need "e" to successfully complete Step 1e and 1f.

You should now have an equation that looks like:
$$\# = \#e^{\#k}$$

1d. Divide both sides by the coefficient in front of $e^{\#k}$. Compute the division on the left to get a new number. You should now have an equation resembling:
$$\# = e^{\#k}$$

1e. Take the natural log (ln) of both sides:
$$\ln\# = \ln e^{\#k}$$

1f. Compute the left side on the calculator to get a number. On the right side, "lne" cancels out leaving #k, as:
$$\# = \#k$$

Divide both sides by the coefficient in front of k to get the value of k:
$$k = \#$$

Now, re-write the formula with the number you just found for k, which should resemble:
$$A = A_0e^{\#t} \text{ or, as a function: } f(t) = A_0e^{\#t}$$

If you know A_0, you can put that in too, as:
$$A = \#e^{\#t} \text{ or as a function: } f(t) = \#e^{\#t}$$

Either form is acceptable. This is the function (formula) you're trying to determine to answer question 1. You will use this for the next part.
Continued on the next page...

2. Find t

2a. Begin by finding the value of "A" by doing a mini-preliminary calculation.

If the problem is a *growth* problem, multiply the fraction or percent (in decimal form) times the *final* amount, which should have been given. You can think of it as:

$$\frac{A_f}{\#} = A_0 e^{\#t}$$

or with the percent in decimal form:

$$(0.\#\#)A_f = A_0 e^{\#t}$$

If the problem is a *decay* problem, multiply the fraction or percent (in decimal form) times the *initial* amount, A_0. It will look like:

$$\frac{A_0}{\#} = A_0 e^{\#t}$$

or with the percent in decimal form:

$$(0.\#\#)A_0 = A_0 e^{\#t}$$

If you must find the *half-life*, or the time for a substance to decay to half its original amount, put

$$\frac{A_0}{2}$$ or $0.5A_0$ in for A, then calculate the number.

2b. Plug the number for amount (that you just found) in for A, and plug the number for the starting amount in for "A_0" (the same you used in Part 1). It will look like:
$$\# = \#e^{\#t}$$

2c. Divide both sides by the number in for A_0, and compute the division of numbers on the left, which will give you a new number on the left as:
$$\# = e^{\#t}$$

2d. Take the natural log (ln) of both sides:
$$\ln\# = \ln e^{\#t}$$

2e. Compute the left side on the calculator to get a number. On the right side, "lne" cancels out leaving #t, as:
$$\# = \#t$$

2f. Divide both sides by the coefficient in front of t to get the value of t:
t = #

Remember, you were likely given a starting year in the problem, and you just found t, but t is not in years, it is the *time from the starting year*.

2g. Add the value of t you just found to the starting year to get the year of the given Amount for the question.

Back to WP53 (pg 212).

Note: Continuous Growth can also be applied to Compounding Interest as seen in WP55 (pg 220).

Logistic Growth

Problems like this, which track the population count as a function of time, are often scenarios involving the spreading of an infectious disease as in the case of a viral epidemic or pandemic. Logistic Growth models are built on the formula:

$$f(t) = A = \frac{c}{1 + ae^{-bt}}$$

where:
a & b are constants specific to the context of the model;

c is also a constant representing the maximum amount, which is why the upper horizontal asymptote is "y = c". (If the context of the problem is population, then c represents the maximum population to be infected by the epidemic);

e = Euler's Number, often called "the number e." This number is also a constant, 2.72, when rounded to the nearest hundredths place, however, the actual number does not need to be used in logistic growth problems because taking the natural log of e cancels it out (as seen in Step 7 of the procedure when solving for t);

t = time in the units of the context of the problem.

It should be understood that these types of problems are based on mathematical *models*, meaning the values of constants a, b and c, and thus the function and graphical representation, come from real, contextual data. The function or model is then used to make accurate estimates of data at various points of time.

A Few Noteworthy Points
• The shape of the graph of the function is known as a Sigmoid Curve, which can simply be thought of as an S-shaped curve with lower and upper limits defined by horizontal asymptotes.

• The **limit of growth** is defined by the **upper horizontal asymptote**, "y = c", so it can be said that growth cannot exceed the value of numerator "c".

• The **lower horizontal asymptote** is always y = 0

- The **point of maximum growth** is the *inflection point* of the graph. The formulas used to find this point are shown on page 103.

- The value of A when an epidemic began at t = 0 is the same as the **y-intercept** of the graphed curve, which can be found using the formula:
$$\text{y intercept} = \frac{c}{1 + a}$$

Typical Questions that could be asked include:
1. How many people became infected with the virus at the beginning of the epidemic? (See previous bullet point, just above).
2. How many people were infected by a certain given time since the epidemic began?
3. At what time will the number of infected people reach a certain given amount?
4. What is the point of maximum growth?
5. What is the limiting amount of people that get infected?

Procedure for Logistic Growth Calculations
It is very important to follow order of operations when simplifying this equation; and the exact procedure you use to simplify depends on what is given and what is being solved for.

When t is given (as in Questions 1 & 2 of WP54):
1. Simplify (multiply) the exponent part of e to get one number as the exponent.
2. Compute $e^{\#}$ to get a number. Remember to properly include the negative sign of the exponent when using your calculator. Round this number to the thousandths place.
3. Multiply this number by the coefficient in front of e. Round the product to the thousandths place.
4. Add the numbers in the denominator.
5. Divide the numerator by the denominator.

When solving for t (as in Question 3 of WP54):
1. Put the entire denominator in parentheses and cross multiply, but do not distribute, for the reason you will see in the next step.
2. Divide both sides by value in for A, which, after Step 1, is the coefficient in front of the parenthetical group.
The value for A should now be in the denominator under the c-value.
3. Divide the numbers on the right side to get a new number. Keep at least 3 decimal places of this number.
Continued on the next page...

4. Subtract the "1" from both sides. This eliminates it on the left and must be subtracted from the number found by division in Step 3.

5. Divide both sides by the coefficient in front of e. This eliminates it on the left, isolating the $e^{\#t}$ term.

6. On the right side, divide the numbers to get a new number. Keep at least 3 decimal places of this number.

7. Take the natural log (ln) of both sides. This will eliminate "lne" on the left, bringing down and leaving the exponent portion. On the right, compute the natural log of the number to get a new number. Keep at least 3 decimal places of this number.

8. Divide both sides by the coefficient in front of "t".
This results in the value of "t". Depending on the units of t, convert the decimal part of the number into the appropriate unit using the appropriate conversion factor.

Finding the Point of Maximum Growth. Use the formulas:

$$x = \frac{\ln(a)}{b} \text{ and } y = \frac{c}{2} \text{ to get } (x, y)$$

Back to WP54 (pg 216).

Compounding Interest

Compounding interest may be calculated using the function/formula:

$$f(t) = A = P\left(1 + \frac{r}{n}\right)^{nt}$$

where
A = total Amount of money with respect to time (Amount is a function of time),
P = the Principle,
r = the annual interest rate as a percent in decimal form,
n = the number of times the money is compounded per year, and
t = time, in years, the money is invested or borrowed for.

If n = 1, you may see the formula as:
$$A = P(1 + r)^t$$

Compounding Interest may also be calculated by:

Compounding Interest as Continuous Growth
This is the same function or formula as the Exponential (Continuous) Growth function (pg 96), just with a couple different symbols to fit the context of money and interest:
$$f(t) = A = Pe^{rt}$$

where
A is the total Amount of money with respect to time (Amount is a function of time),
P, in place of A_0, is the Principle,
e = Euler's Number, often called "the number e",
r, in place of k, is the annual interest rate as a percent in decimal form, and
t is time, in years.

Back to WP55 (pg 220).

IDENTIFY & MATCH PRACTICE

This section is to help you with the first step of solving word problems, which is *Identification*. If and when you can *identify the type* of word problem you encounter, you automatically increase your chance of being able to set up the equation correctly. Read each word problem and look for key clues, then determine the *type* of problem it is, by words and name. Analyze the problem by asking the questions from page 34 to help you identify the problem as specifically as you can, to determine its type or category. Then use your answers to choose from The Categories (pg 39).

The answers, as well as page numbers to the fully setup and worked out problems, can be found on page 107. Consider writing down your answers before going to the Answer page.

I&M A: In the previous year, Ian of Ian's Pumpkin Carvings made $73,510. The next year, sales were up, giving him a 9% salary increase. How much did Ian make the next year (after the 9% increase)?

I&M B: Carl approached a toll-booth which charges $1.25. In his change compartment, he only had nickels and dimes, and had the exact change needed to pay the toll. If his number of nickels was four more than five times the number of dimes, how many nickels and dimes did he have?

I&M C: A rectangle, whose width is three cm less than four times its length, has an area of 175 cm^2. Determine the length and width.

I&M D: There has been a bank robbery in the middle of the night in Columbus, Ohio. The robbers took off in a white van headed west on I-70 driving at 85 miles per hour. Police discovered the robbery exactly one hour after the robbery took place and immediately took off west on I-70 in an unmarked car, driving at 100 miles per hour. When the police caught up to the robber's van, they immediately pulled them over and apprehended the suspects. How long did it take the police to catch the bank robbers?

I&M E: Ed invested $300 in an investment which pays quarterly dividends at a 15% interest rate. What was the interest gained in the first quarter?

I&M F: It takes Alison 3 hours to complete a task if she works alone, and it takes Claire 4 hours to complete the same task if she works alone. How long will it take to complete the task if they work together?

I&M G: Michele painted a square picture and wants to frame it in a larger, square frame. If a side of the frame is 3 cm less than double the size of a side of the painting, and the area of the painting is 24 cm^2 less than the whole area inside the frame, what are the dimensions of the painting and the frame? What is the area of the painting, what is the area of the whole framed painting, and what is the area of just the frame?

I&M H: A company manufactures microphones and speakers for smart-phones. Each microphone costs $6 and takes 3 minutes to make. Each speaker costs $7 and takes 5 minutes to make. If a total of $1259 was spent in a total of 772 minutes to make a batch of these parts, how many of each part was made?

I&M I: A triangle in which the first side is one cm less than twice the length of the second side, and the third side is one cm more than the second side, has a perimeter of 20 cm. What are the lengths of each side?

I&M J: Greg has a pocket full of change consisting only of nickels, dimes, and quarters. The total value of the change is $3.35 and there are a total of 25 coins. He has ten more dimes than nickels. How many of each coin does he have?

I&M K: If the sum of three consecutive odd integers is 33, find the integers.

I&M L: Mark flew 2100 miles from Philadelphia to Las Vegas to attend a bachelor party, and then flew back 3 days later. Flying westward against the jet stream, the flight took 300 minutes. Flying home, eastward, with the jet stream, the flight took 270 minutes. Based on this information, what was the speed of the plane flying west, against the jet stream? What was the speed of the plane flying east, with the jet stream? What was the speed of the jet stream? And what would the speed of the plane be in still-air (no jet stream/air current)?

I&M Answers

A. Salary/Percent Increase, New Salary Unknown; pg 133.

B. A Mixed Items problem involving Two Coins of Different Value solved with two unknowns, one variable and one equation, in which one unknown is in-reference-to the variable; pg 113.

C. Area of a Rectangle with two unknowns, one variable and one equation, in which one unknown is in-reference-to the variable; pg 177.

D. A Rate of Speed related Catch-Up problem in which the times of two vehicles or travelers are unknown. One unknown will be in-reference-to the variable. The given rates of speed and the difference in the times are used to solve; pg 145.

E. A Simple Interest Investment Problem in which one investment is made and the interest gained is unknown. The variable is the only unknown; pg 139.

F. A Splitting A Task problem in which the time to complete a task together is unknown; pg 164.

G. A problem involving the Areas of Two Squares, which sets up a quadratic formula and can be solved by either factoring or the quadratic formula. One unknown will be in-reference-to the variable; pg 181.

H. A Manufacturing of Two Different Items problem with two variable which must be solved as a system of two linear equations; pg 197.

I. A Perimeter problem involving a triangle, in which the unknown lengths of two sides are in-reference-to the variable; pg 175.

J. A problem with three unknowns, specifically involving three differently valued coins, which must be solved using a system of three linear equations; pg 209.

K. A Sum of Three Consecutive Odd Integers problem (yes, this is fairly obvious); pg 111.

L. A Rate of Speed (Upstream/Downstream) problem involving the rate of the current and the rate of the vehicle in no current, which must be solved using a system of two linear equations with two variables; pg 158.

ANNOTATED EXAMPLES

WP1: An Unknown Among an Average

The Problem

Emily needs a test score average of 90 in her algebra class to graduate with honors. There are 5 tests of equal weight. On the first 4 tests, she received grades of 91, 93, 97, and 70. What must she score on the fifth test so her final test score average is 90?

Identify

This is a problem involving averages. The missing data point (test grade) is the only unknown, and will be assigned as variable x. Note: There is no additional *detailed information* on this problem, as "calculating average" is a very basic formula and equation (shown below).

The Unknown

Let x = the grade of the fifth test

The Formula for Average:

$$\frac{\text{Sum of all data points} + x}{\text{Total \# of data points}} = \text{Average}$$

The Setup

$$\frac{91 + 93 + 97 + 70 + x}{5} = 90$$

The Math

Add the like-terms (the four numbers) in the numerator, then cross multiply (think of 90 as being over 1) to get:
$351 + x = 450$

Subtract 351 from both sides, and

$x = 99$

The Solution

Emily must get a score of 99 on the fifth test to achieve a final test score average of 90.

WP2: Consecutive Integers

The Problem

If the sum of three consecutive integers is 45, find the integers.

Identify

This is a consecutive integers problem with three unknowns (pg 45). One of the unknowns is variable "n"; the other two unknowns are in-reference-to variable n.

The Unknowns

Let $n = 1^{st}$ integer
Let $(n + 1) = 2^{nd}$ integer
Let $(n + 2) = 3^{rd}$ integer

The Template

$n + (n + 1) + (n + 2) = \#$

The Setup

$n + (n + 1) + (n + 2) = 45$

The Math

Remove the parentheses and combine like-terms to get:
$3n + 3 = 45$

Subtract 3 from both sides to get:
$3n = 42$

Divide both sides by coefficient 3:
$$\frac{3n}{3} = \frac{42}{3}$$

$n = 14$

Substitute 14 in for n in:
2nd integer $= n + 1 = 14 + 1 = 15$
and
3rd integer $= n + 2 = 14 + 2 = 16$

The Solutions

1st integer $= 14$
2nd integer $= 15$
3rd integer $= 16$

WP3: Consecutive Odd Integers

The Problem

If the sum of three consecutive odd integers is 33, find the integers.

Identify

This is a consecutive odd integers problem with three unknowns (pg 46). One of the unknowns is variable "n"; the other two unknowns are in-reference-to variable n.

The Unknowns

Let $n = 1^{st}$ integer
Let $(n + 2) = 2^{nd}$ integer
Let $(n + 4) = 3^{rd}$ integer

The Template

$n + (n + 2) + (n + 4) = \#$

The Setup

$n + (n + 2) + (n + 4) = 33$

The Math

Combine like-terms:
$3n + 6 = 33$

Subtract 6 from both sides to get:
$3n = 27$

Divide both sides by coefficient 3:

$$\frac{3n}{3} = \frac{27}{3}$$

$n = 9$

Substitute 9 in for n in:
2nd integer $= n + 2 = 9 + 2 = 11$
and
3rd integer $= n + 4 = 9 + 4 = 13$

The Solutions

1st integer $= 9$
2nd integer $= 11$
3rd integer $= 13$

WP4: Consecutive Even Integers

The Problem

If the sum of two consecutive even integers is 22, find the integers.

Identify

This is a consecutive even integers problem with only two unknowns (pg 46). One of the unknowns is variable "n"; the other unknown is in-reference-to variable n.

The Unknowns

Let $n = 1^{st}$ integer
Let $(n + 2) = 2^{nd}$ integer

The Template

$n + (n + 2) = \#$

The Setup

$n + (n + 2) = 22$

The Math

Combine like-terms to get:
$2n + 2 = 22$

Subtract 2 from both sides to get:
$2n = 20$

Divide both sides by coefficient 2:
$$\frac{2n}{2} = \frac{20}{2}$$

$n = 10$

Substitute 10 in for n in:
2nd integer $= n + 2 = 10 + 2 = 12$

The Solutions

1st integer $= 10$
2nd integer $= 12$

MIXED ITEMS USING ONE VARIABLE & ONE EQUATION

WP5: Two Coins of Different Value (One Variable)

The Problem
Carl approached a toll-booth which charges $1.25. In his change compartment, he only had nickels and dimes, and had the exact change needed to pay the toll. If his number of nickels was four more than five times the number of dimes, how many nickels and dimes did he have?

Identify
This is a mixed items problem (pg 83) involving two coins of different value with two unknowns (pg 85). One unknown will be the variable and the other will be in-reference-to the variable.

The Unknowns
Let x = the number of dimes
Let $5x + 4$ = the number of nickels

Note: You could also let d = the number of dimes and "$(5d + 4)$" = number of nickels.

The Template
(value of coin 1)(# of coins 1) + (value of coin 2)(# of coins 2) = value of all coins

$0.\#\#(x) + 0.\#\#(\#x +/- \#) = \#.\#\#$

Be sure to use the value of each coin in dollar (decimal) form (not as cents).

The Setup
$0.10x + 0.05(5x + 4) = 1.25$

The Math
Distribute the 0.05 through the parentheses to get:
$0.10x + 0.25x + 0.20 = 1.25$

Move the 0.20 to the right by subtracting it from both sides, then combine like terms to get:
$0.35x = 1.05$

Divide both sides by coefficient 0.35:
$$\frac{0.35x}{0.35} = \frac{1.05}{0.35}$$

$x = 3$

Plug 3 in for x in:

The number of nickels = $5x + 4 = 5(3) + 4 = 19$

The Solutions
There were 19 nickels and 3 dimes.

See Also: Two Coins of Different Values solved with two variables and two equations (pg 193).

WP6: Three Coins: A Reference-To-A-Reference-To One Variable

The Problem
Rafael has a pocketful of change with a total value of $2.80. If he has three more quarters than twice as many dimes, and two less dimes than nickels, how many of each coin does he have?

Identify
This is a problem involving three coins, representing a problem with three unknowns where one unknown is the variable, the second unknown is in-direct-reference-to the variable, and the third unknown is in-reference-to the second (intermediate) unknown (pg 14). The equation will be built by adding all three unknowns and will then proceed by distribution, then combining like-terms.

Notice, specifically in this problem, that quarters references dimes which references nickels. Nickels is the root-reference, so let x = number of nickels, then build the unknown for number of dimes with respect to x, then build the unknown for number quarters based on the number of dimes.

The Unknowns
Let x = # of nickels
Let $(x - 2)$ = # of dimes
# of Quarters: $[2(\text{# of dimes}) + 3]$
Substitute "$(x - 2)$" in for "# of dimes":
Let $[2(x - 2) + 3]$ = # of quarters

The Template
(value of coin$_1$)(# of coins$_1$) + (value of coin$_2$)(# of coins$_2$) + (value of coin$_3$)(# of coins$_3$) = total value of all coins

$0.\#\#x + 0.\#\#(x +/- \text{compensation #}) + 0.\#\#[$ (unknown 2) $+/-$ compensation w/ respect to unknown 2] $= \#.\#\#$

The Setup
$0.05x + 0.10(x - 2) + 0.25[2(x - 2) + 3] = 2.80$

115

The Math
Simplify inside the brackets first by distributing the 2 through the "(x − 2)":
$$0.05x + 0.10(x - 2) + 0.25[2x - 4 + 3] = 2.80$$

Combine like-terms inside the brackets:
$$0.05x + 0.10(x - 2) + 0.25[2x - 1] = 2.80$$

Distribute the "0.10" through "(x − 2)" and distribute "0.25" through "[2x − 1]" to get:
$$0.05x + 0.10x - 0.20 + 0.50x - 0.25 = 2.80$$

Combine like-terms on the left to get:
$$0.65x - 0.45 = 2.80$$

Add 0.45 to both sides to get:
$$0.65x = 3.25$$

Divide both sides by coefficient 0.65:
$$\frac{0.65x}{0.65} = \frac{3.25}{0.65}$$

$$x = 5$$

Substitute 5 in for x in:
# of dimes = (x − 2) = 5 − 2 = 3

Substitute 5 in for x in:
# of quarters = [2(x − 2) + 3] = [2(5 − 2) + 3] = 9

Note: You could have also substituted "3" in for "# of dimes," (x − 2), in the "# of quarters" mini-equation.

The Solutions
Rafael has 5 nickels, 3 dimes and 9 quarters.

WP7: Two Different Priced Tickets (One Variable)

The Problem
A famous comedian performed a stand-up set at a local club. Tickets for front row seats were $60, tickets for regular seats were $30, and brought in a total of $1890. If the club sold out with 52 tickets, how many seats of each type does the club have?

Identify
This is a mixed items problem (pg 83) involving tickets with two different prices (pg 87). There are two unknowns. One variable will represent one unknown, and the other unknown will be in-reference-to that variable, both in one equation. (See Note on pg 118).

The Unknowns
Let x = the number of front row seat tickets
Let $(52 - x)$ = the number of regular seat tickets

The Template
(ticket price$_1$)(# of tickets$_1$) + (ticket price$_2$) (total tickets $- x$) = total revenue

$$\#x + \#(\# - x) = \#$$

The Setup
$60x + 30(52 - x) = 1890$

The Math
Distribute the 30 through the parentheses to get:
$60x + 1560 - 30x = 1890$

Move the 1560 to the right by subtracting it from both sides, then combine like-terms:
$30x = 330$

Divide both sides by coefficient 30:
$$\frac{30x}{30} = \frac{330}{30}$$

$x = 11$

Substitute 11 in for x into:
The number of regular seats = 52 – x = 52 – 11 = 41

The Solutions
There are 11 front row seats and 41 regular seats.

Note: Since *two* totals are given, this problem could also be solved with two variables and two equations as a system of two liner equations as in WP46 (pg 195).

WP8: Buying Mixed Items at Two Different Unit Prices (One Variable)

The Problem
Katrina went into the Jersey Shore Boardwalk Candy Shop and spent $21.74 (not including sales tax) on a mixed bag of salt water taffy and fudge. The salt water taffy costs $8.49/lb and fudge costs $11.99/lb. If she bought a total of two and a quarter pounds, how many pounds of each candy did she buy?

Identify
This is a *mixed items with different unit prices* problem (pg 88). One unknown will be the variable and the other unknown is in-reference-to the variable and the total, since the total number of pounds is given.

The Unknowns
Let x = the weight of salt water taffy purchased at $8.49/lb
Let (2.25 – x) = the weight of the fudge purchased at $11.99/lb

The Template
(unit price$_1$)x + (unit price$_2$)(total weight – x) = total dollars spent
or
#.##x + #.##(# - x) = #.##

The Setup
Convert "two and a quarter pounds" into numeric decimal form 2.25 before putting it in as the total weight:

8.49x + 11.99(2.25 – x) = 21.74

The Math
Distribute the 11.99 into the (2.25 – x) to get:

8.49x + 26.98 – 11.99x = 21.74

Combine like-terms on the left. Then, move the 26.98 to the right by subtracting it from both sides. This will give:
-3.5x = -5.24

x = 11

Divide both sides by coefficient "-3.5":

$$\frac{-3.5x}{-3.5} = \frac{-5.24}{-3.5}$$

x = 1.5

Substitute 1.5 in for x into:
The weight of fudge purchased = 2.25 – x = 2.25 – 1.5 = 0.75

The Solutions
Katrina purchased 1.5 lb of salt water taffy and 0.75 lb of fudge.

Note: This problem can also be solved as a system of two linear equations as in WP50 (pg 203).

WP9: Mixing Two Chemicals to Make a Final Solution (One Variable)

The Problem
Nina works in the fragrance industry and was developing a perfume. To get the scent just right, she needed to make a final volume of 2 Liters of a 22% solution made from a mixture of a 3% solution and a 27% solution. What volume of each solution must she use to achieve her desired final solution?

Identify
This is a mix problem about mixing two chemicals of different percentage concentrations to make a final mixture of a defined percentage concentration (pg 89) and will be solved using one variable and one equation. (See Note on pg 122).

The Unknowns
Let x = the volume of 3% solution
Let (2 − x) = the volume of 27% solution. This is in reference to x.

Note: You could assign these unknowns to the opposite percentages, and get the same solution.

The Template
(% in dec. form)(vol)$_1$ + (% in dec. form)(total vol - x)$_2$ = (% in dec. form)(total vol.)$_{final}$

$$0.\#\#x + 0.\#\#(\# − x) = (0.\#\#)(\#)$$

The Setup
Convert each percentage into decimal form by moving the decimal two places to the left. This is especially important to remember for single digit percents.
Substitute the decimal-form numbers into the template equation to get:
$$0.03x + 0.27(2 − x) = 0.22(2)$$

The Math
On the left side, distribute the 0.27 through the parentheses. On the right side, multiply the two numbers to get a new number:
$$0.03x + 0.54 − 0.27x = 0.44$$

On the left, combine the x-terms:
-0.24x + 0.54 = 0.44

Move the 0.54 to the right side by subtracting it from both sides, then combine like-terms on the right, to get:
-0.24x = -0.10

Divide both sides by coefficient -0.24:
$$\frac{-0.24x}{-0.24} = \frac{-0.10}{-0.24}$$

x = 0.42, when rounded to the hundredths place.

To find the volume of 27% solution needed, substitute the value for x just found, 0.42, into
(2 − x): 2 − 0.42 = 1.58

The Solutions
0.42 Liters of 3% solution and
1.58 Liters of 27% solution were needed.

Note: This problem can also be solved using two variables and two equations as in WP48 (pg 199).

MONEY RELATED
WP10: Fees & Membership Costs = Total Bill

The Problem
Donna is planning her budget, which involves predicting her upcoming cell phone bill. Her monthly fee is $65.24, not including data overage charges and app purchases. If data overage charges are 6 cents per megabyte, apps are $2.25 each, and she purchased 3 apps, how many megabytes did she go over if her total bill was $116.09?

Identify
This is a problem in which fees and membership costs make up the total bill (pg 53). There is one unknown, which will be the variable.

The Unknown
Let x = the number of overage megabytes

The Template
initial fee + (# of units)(cost per unit) + other charges = total bill

#.## + (#.##)(#) + #.##x = #.##

The Setup
Convert the 6 cents to dollars as 0.06.

65.24 + (2.25)(3) + 0.06x = 116.09

The Math
Multiply the "(2.25)(3)" to get:
65.24 + 6.75 + 0.06x = 116.09

Combine like-terms on the left to get:
71.99 + 0.06x = 116.09

Move the 71.99 to the right by subtracting it from both sides to get:
0.06x = 44.10

Divide both sides by coefficient 0.06:
$$\frac{0.06x}{0.06} = \frac{116.09}{0.06} ; x = 735$$

The Solution: Donna went over by 735 megabytes.
123

WP11: Expenses & Profit

The Problem
In a typical week, Napoli's Italian Restaurant makes an average of $2789.23 in gross profit, and weekly expenses total $1,414.35. Due to an increase in customers, the owner would like to have an addition built on to the restaurant to seat more customers. An estimate suggests the addition will cost $22,000. How many weeks of net profit will it take to earn the amount needed for the addition?

Identify
This is an *expenses & profit* (pg 54) problem where the number of weeks is the only unknown, and will be the variable.

The Unknown
Let x = number of weeks

The Template
(Gross Profit)x – (Expenses)x = Target Net Profit

#x - #x = #

The Setup
2789.23x – 1414.35x = 22000

The Math
Combine like-terms:
1374.88x = 22000

Divide both sides by coefficient 1374.88:
$$\frac{1374.88x}{1374.88} = \frac{22000}{1374.88}$$

x = 16 (rounded to the nearest whole number)

The Solution
It will take 16 weeks to earn enough profit to pay for the addition.

WP12: Expenses & Profit (More Complicated)

The Problem

An up-and-coming band has five members and plays three shows per week. All members have a Monday through Friday day-job, but would prefer to be full-time musicians. Each member currently makes a $50,000 annual salary from the day-job and would quit the day-job if and when each made a yearly net profit from playing gigs equal to their day-job salary. They consistently sell 61 tickets at $25 per ticket, per show. Each show costs the band $20 in expenses and each week costs $10 in rental fees. How many weeks must they play this schedule in order to make enough to quit their day jobs? Or will they not make enough to be able to quit their day jobs?

Identify

This is a more complicated *expenses & profit* problem (pg 55). The equation will be set up to subtract expenses from gross profit to equal net profit. There is only one unknown, which will be the variable, but it will appear in more than one place. Some of the given information will need to be used in preliminary mini-calculations to properly set up and fill-in the main equation. Note: Although tickets are involved in this example, this is *not* a "mixed items: two different priced ticket" problem.

The Unknown

Let x = number of weeks
The only unknown is the variable x, but it is used twice in the following equation.

The Template

Gross Profit − Expenses = Net Profit

Use a *mini-equation* to determine the Gross Profit per week from three shows:

$$\text{Weekly Gross Profit} = \left(\frac{\text{\# tickets}}{\text{day}}\right)\left(\frac{\$}{\text{ticket}}\right)\left(\frac{\text{days}}{\text{week}}\right)$$

where "tickets" cancels "ticket" and "day" cancels "days" leaving $ per week.

Use a *mini-equation* to determine Total Weekly Expenses:

$$\text{Total Weekly Expenses} = \left(\frac{\text{expenses}}{\text{day}}\right)\left(\frac{\text{days}}{\text{week}}\right) + \left(\frac{\text{other expenses}}{\text{week}}\right)$$

where "day" cancels "days" creating like-fractions which can be added to give $ per week.

Use a mini-calculation (where "member" cancels with "members") to determine the Net Profit the band must make in 1 year:

$$\text{Net Profit Needed} = \left(\frac{\text{salary}}{\text{member}}\right)(\text{\# of members})$$

Substitute the given numbers from the preliminary mini-equations into the **template equation**:

#x - #x = #

The Setup
Preliminary steps can be done using the mini-equations above before setting up the main equation.

Weekly Gross Profit = (61)(25)(3) = 4575

Total Weekly Expenses = (20)(3) + 10 = 70

Net Profit Needed = (50000)(5) = 225000

Put these numbers into the template equation to make:
4575x − 70x = 225000

The Math
Combine like-terms by performing the subtraction to get:
4505x = 225000

Divide both sides by coefficient 4505:

$$\frac{4505x}{4505} = \frac{225000}{4505}$$

x = 49.945, rounded to 50 since you can't have a fraction of a show.

The Solution
Since the band members will earn enough, they can quit their day-jobs after 50 consecutive weeks of gigs.

PERCENT

WP13: Percent: Part of Whole Unknown

The Problem
What is 25% of 12?
Can also be asked as: 25% of 12 is what number?

Identify
This is a percentage problem where the "part" of the *whole* or *original amount* is unknown (pg 48). As this is the only unknown, this will be the variable.

The Unknowns
Let x = the part of the whole or original amount

The Template
(% in decimal form)(original amount) = x

$(0.\#\#)(\#) = x$

The Setup
Convert 25% to decimal form by dividing 25 by 100 or moving the decimal two places to the left.

$(0.25)(12) = x$

The Math & The Solution
Multiply; this is the only step.
x = 3

WP14: Percent: Size of Original Amount Unknown

The Problem
At a certain school, the top 25% of the class graduate with honors status. If three hundred students graduated with honors status, what was the size of the class?

Identify
The question could be simplified as: Three hundred is 25% of what number? This is a percentage problem in which the *original amount* or *whole* is unknown (pg 48).

The Unknown
x = the original amount

The Template
(percent in decimal form)x = part of the original amount

0.##x = #

The Setup
Convert 25% to decimal form by dividing 25 by 100, or moving the decimal two places to the left.

0.25x = 300

See alternative setup using proportion on the next page.

The Math
Divide both sides by coefficient 0.25:
$$\frac{0.25x}{0.25} = \frac{300}{0.25}$$

x = 1200

The Solution
The total number of students in the graduating class was 1200 students.

Alternative Setup Using Proportion:

$$\frac{300}{x} = \frac{25}{100}$$

The Math

Cross multiply to get:

$25x = 30000$

Divide both sides by coefficient 25:

$$\frac{25x}{25} = \frac{30000}{25}$$

$x = 1200$

The Solution: 1200 students

WP15: Percent: Percent Unknown

The Problem
The United States House of Representatives has 435 seats. As part of the process for a law to be made, it starts as a bill. The bill needs to pass with a *simple majority*, meaning "over 50% in favor" by the House before moving to the Senate. If a new bill gets 219 *in favor* votes, will it win the simple majority to move to the Senate?

Identify
This question could be simplified as: What percent of 435 is 219?
It could also be asked as: 219 is what percent of 435?
This is a percentage problem in which the percent (in decimal form) is unknown (pg 48).

The Unknowns
Let x = the percent in *decimal* form

Once x is found:
(100%)(x) = the answer to the question in *percent* form

The Template
You could write:
x(original amount) = part of the original
but the variable is usually written *after* the number, so it is more likely to look like:
(original amount)x = part of original

#x = #

The Setup
435x = 219

See alternative setup using proportion on the next page.

The Math
Divide both sides by coefficient 345
$$\frac{435x}{435} = \frac{219}{435}$$

x = 0.503

This is currently in decimal form and needs to be converted to a percent by multiplying

$(100\%)(0.503) = 50.3\%$

The Solution
Yes, the bill will move to the Senate because it passed in the House with 50.3%.

Alternative Setup Using Proportion:
$$\frac{219}{435} = \frac{x}{100}$$

Cross-multiplied gives:

$435x = 21900$

Divide both sides by coefficient 435:
$$\frac{435x}{435} = \frac{21900}{435}$$

$x = 50.3$

The Solution
When performed in proportion form, the answer comes out directly as a percentage with no need to convert from decimal form. Yes, the bill will move to the Senate because it passed in the House with 50.3%.

WP16: Salary/Percent Increase: New Salary Unknown

The Problem
In the previous year, Ian of Ian's Pumpkin Carvings made $73,510. The next year, sales were up, giving him a 9% salary increase. How much did Ian make the next year (after the 9% increase)?

Identify
This is a percent increase (in regards to salary) problem where the *new amount* is unknown (pg 50). Since this is the only unknown, the variable will represent that unknown.

The Unknown
Let x = Ian's salary next year

The Template
(previous salary) + (% in dec. form)(previous salary) = new, higher salary

+ (0.##)(#) = x

The Setup
Convert 9% into decimal form, 0.09, and put it into the template equation:
73510.00 + (0.09)(73510.00) = x

The Math
Multiply 0.09 times the previous salary, 73510.00:
73510.00 + 6615.90 = x

Add the numbers on the left to get:

80125.90 = x

The Solutions
Ian's salary next year is $80,125.90

WP17: Percent Increase (Sales Tax): Original Amount Unknown

The Problem
Suppose you go Christmas shopping with the intention of buying gifts for three people, using only the cash you have, which is $357.00. If you spend the same amount of money on each person's gifts, and sales tax is 7%, what is the most you could spend on each person, *before tax*?

Identify
Although you may not realize it at first, because of the context, this is a percent increase problem where the original amount is unknown (pg 51). The increase taking place is the pre-tax price to the actual, tax-included price one must pay. There is only one unknown so that will be the variable, but you must divide that amount by 3 at the end to answer the question.

The Unknowns
Let x = the total amount of money spent on all three people, before tax
Let $\dfrac{x}{3}$ = the amount you can spend on each person

The Template
(original amount) + (percent in decimal form)(original amount) = the new, higher amount

$x + 0.\#\#x = \#$

The Setup
$x + 0.07x = 357.00$

Note: You could also set it up as: $x = 357.00 - 0.07x$
by thinking of it as "the original, non-tax price equals the tax-included price minus the tax on the original price."

The Math
Combine like-terms (treating x as 1x) to get:
$1.07x = 357.00$

Continued on the next page...

$1.07x = 357.00$

Divide both sides by coefficient 1.07:
$$\frac{1.07x}{1.07} = \frac{357.00}{1.07}$$

$x = 333.64$ (rounded to the nearest cent, the hundredths place), which is the total amount you can spend before tax is included, but the question asks how much you can spend per person, pre-tax. Since there are three people to buy for, divide $333.64 by 3, which is:

$111.21 (rounded to the nearest cent)

The Solution
You can spend $111.21 per person.

The *Wrong Approach* is to take 7% of $357, subtract it from $357, then divide that amount by 3. This is wrong because this is taking 7% of the final amount, and **percent is not the same in the forward direction as it is in the reverse direction.** An example of this is explained in Percent Decrease (pg 52).

WP18: Salary/Percent Increase: Percent Unknown

The Problem
Last year, Melyssa, an environmental scientist, was making a salary of $54,000. If she now makes $57,000, what percent raise did she receive from last year to this year?

Identify
This is a percent increase problem where the "percent" is unknown (pg 51). Since there is only one unknown, the variable will be the unknown.

The Unknown
Let x = the percent increase, in decimal form

The Template
(original salary) + (% dec. form)(original salary) = new, higher salary

+ #x =

Note: Notice that "% in dec. form" is shown to the left of the factor "original salary" it is multiplied by. In the setup equation, however, "x", which represents the unknown percent in decimal form, is shown *after* the original salary, 54000. They are presented in this order because the variable tends to be written after the coefficient.

The Setup: $54000 + 54000x = 57000$

The Math
Subtract "54000" from both sides to combine constants on the right and isolate the term with the variable on the left, which becomes:

$54000x = 3000$

Divide by coefficient 54000: $\dfrac{54000x}{54000} = \dfrac{3000}{54000}$

$x = 0.056$, rounded to the thousandths place

Since x was assigned as "the percent as a decimal," that's what you found, 0.056, however, that is not the final answer because that is not in percent form. To **convert** it to percent form, multiply 0.056 times 100% (which moves the decimal two places to the right).
The Solution: Melyssa received a 5.6% raise.

WP19: Percent Decrease

The Problem
Nicoletta bought a new car in June of 2013 at book value. Exactly one year later, in 2014, she decided she wanted to sell it. Since it was purchased, the car depreciated by 20%, as many new cars do in the first year, and was worth $19,450.00 in 2014. What was the original price of the car?

Identify
This is a percent decrease problem (pg 52). Since the original amount is the only unknown, it will be represented by the variable. Because of the type of problem and equation, the variable will appear in two places in the equation.

The Template
original amount – (% in dec. form)(original amount) = new, decreased amount

$$x - 0.\#\#x = \#$$

The Setup
$$x - 0.20x = 19450.00$$

The Math
Combine like-terms (treating the x as 1x) to get:
$$0.80x = 19450.00$$

The coefficient 0.80 came from "1 – 0.20". Divide both sides by coefficient 0.80:

$$\frac{0.80x}{0.80} = \frac{19450.00}{0.80}$$

$$x = 24312.50$$

The Solution
The car was purchased for $24,312.50 in 2013.

Note: The **wrong approach** would be to take 20% of 19450 and subtract it from 19450.

INVESTMENTS & LOANS

WP20: Simple Interest: Interest Unknown

The Problem
Ed invested $300 in an investment which pays quarterly dividends at a 15% interest rate. What was the interest gained in the first quarter?

Identify
This is a simple interest (pg 56) problem in which the dividend cycle is quarterly. There is only one unknown so it will be represented by the variable.

The Unknown
The unknown is the value of *interest* which is already assigned as variable "I"

The Formula
$I = Prt$

The Setup & The Math
Convert the interest rate to decimal form by dividing it by 100. Since t (time) is in units of *years*, convert *quarterly* to years as $\frac{1}{4}$ or 0.25:

$I = (300)(0.15)(0.25)$

Multiply.

$I = 11.25$

The Solution
Ed earned $11.25 interest on his investment.

WP21: Simple Interest: Total Amount after Interest Unknown

The Problem
Ed invested $9035 in a semiannual investment with a 7.5% interest rate. What's the total amount of money he will have after the earned interest from the first period is paid to him?

Identify
This is a simple investment problem in which *total amount* is the unknown (pg 60).

The Unknown
You are trying to find the value of A, which is the designated variable for "new amount."

The Formula
$A = P + Prt$

The Setup
Plug the givens in for P, r, & t. Since the period is semiannual, use $\frac{1}{2}$ or the decimal form, 0.5, for t.

$A = 9035 + (9035)(0.075)(0.5)$

The Math
Multiply the factors on the right to get:
$A = 9035 + 338.81$

then add, to get:
$A = 9373.81$

The Solution
Ed has a total, new amount of $9373.81 after the first period.

WP22: Investing in Two Simple Interest Investments (One Variable)

The Problem
Ed received a $5000 rebate from his federal tax return and wanted to invest that money. In order to diversify his portfolio, he put his money in two different investments: part in a simple interest certificate of deposit (CD) and the remaining part in stock in a technology company. In the first year, they earned a combined amount of $462.50. If the CD had an interest rate of 12%, and the stock went up by 7%, how much money did he invest in the CD and how much did he invest in the stock, respectively?

Identify
This problem involves investing in two different investments at different interest rates (pg 90) and falls under the category of "mixed items" (pg 83). Based on the givens, it can be solved using one variable and another unknown in reference-to-the variable, in one equation. (See Note below).

The Template
(% in dec. form)x + (% in dec. form)(total amount *invested* – x) = total Interest earned

$0.\#\#x + 0.\#\#(\# - x) = \#$

The Unknowns
Let x = the amount invested in the CD at 12%
Let (5000 – x) = the remaining amount of money invested in the stock, up 7%

Note: Alternatively, you could have let the amount invested at 7% be "x" and the amount invested at 12% be "(5000 – x)", and the solutions would come out correctly.

The Setup
Convert the percentages to decimal form, then plug them into the template equation:
$0.12x + 0.07(5000 - x) = 462.50$

The Math
Distribute the 0.07 through the parentheses to get:
$0.12x + 350 - 0.07x = 275$

Move the 350 to the right by subtracting it from both sides. Then combine like-terms to get:
$0.05x = 112.50$

Divide both sides by coefficient "0.05":
$$\frac{0.05x}{0.05} = \frac{112.50}{0.05}$$

$x = \$2250$

Substitute $2250 in for x into the other unknown:
The amount invested at 7% = $5000 - x = 5000 - 2250 = 2750$

The Solutions
Ed invested $2250 into the 12% simple interest certificate of deposit and $2750 into the stock which earned 7%.

Note: Since two total amounts are given, this problem can also be solved with two variables and two equations as in WP49 (pg 201).

RATE OF SPEED

WP23: Two Trains Leave the Station, When Will They Meet? Time Unknown, Total Distance Given, Different Speeds

The Problem
Two trains leave their stations at the same time, travelling towards each other at different speeds, 100 miles apart. One train leaves Grand Central Terminal in Manhattan, NY, and the other leaves 30th Street Station in Philadelphia, PA. The Philadelphia-bound train travels 67 miles per hour and the Manhattan-bound train travels 74 miles per hour. If each departed at exactly 12:00 PM, at what time will they pass each other?

Identify
This is a rate of speed problem in which two vehicles are travelling towards each other and time is unknown (pg 71). There is only one unknown, which is time.

The Unknown
Let t = the time it takes the two trains to meet and pass, in hours

Note: time is in hours because the rate is miles per *hour*.

The Template
$r_1t + r_2t = d_{total}$

#t + #t = #

The Setup
67t + 74t = 100

The Math
Combine like-terms to get:
141t = 100

Divide both sides by coefficient 141:
$$\frac{141t}{141} = \frac{100}{141}$$

t = 0.71

It takes the two trains 0.71 hours until they pass each other. The question asks *what time* they will meet if they both left at 12:00 PM. Since clock-time is in hours: minutes, you must **convert** the 0.71 hours to minutes:

$$0.71 \text{ hr} \left(\frac{60 \text{ min}}{1 \text{ hr}} \right) = 42.6 \text{ min, rounded to } 43 \text{ min}$$

then add the minutes to 12:00:
12:00 + 0:43 = 12:43

The Solution
The trains will pass at 12:43 PM

WP24: Catch-Up: Times Unknown, Same Distance, Different Speeds

The Problem
There has been a bank robbery in the middle of the night in Columbus, Ohio. The robbers took off in a white van headed west on I-70 driving at 85 miles per hour. Police discovered the robbery exactly one hour after the robbery took place and immediately took off west on I-70 in an unmarked car, driving at 100 miles per hour. When the police caught up to the robber's van, they immediately pulled them over and apprehended the suspects. How long did it take the police to catch the bank robbers?

Identify
This is a rate of speed problem where the vehicles are going in the same direction and one catches up to the other (pg 72). Since there are two unknowns, one will be the variable and the other will be in-reference-to the variable.

The Unknowns
Since t already represents time,
Let t = the time of the robber's vehicle (that left first)
Let (t – 1) = the time of the police vehicle (that left last)

The Template
$r_1t = r_2(t$ +/- time compensation #)

#t = #(t +/- #)

The Setup
$85t = 100(t - 1)$

The Math
Distribute the 100 through the parentheses to get:
$85t = 100t - 100$

Combine like-terms by moving the "100t" to the left by subtracting it from both sides, to become:
$-15t = -100$

145

Divide both sides by coefficient "-185" to get:

$$\frac{-15t}{-15} = \frac{-100}{-15}$$

$t = 6\dfrac{2}{3}$ as a mixed number, or $\dfrac{20}{3}$ as an improper fraction, or

$t = 6.7$, rounded to the nearest tenths place.

Since we don't really give hours in fraction or decimal form, you might consider **converting** the fractional or decimal portion hours to minutes by the following conversion:

$$\frac{2}{3}\,\text{hr}\left(\frac{60 \text{ min}}{1 \text{ hr}}\right) = 40 \text{ min}$$

The fraction was used instead of the decimal because the fraction form is a more exact value.

$t = 6$ hours and 40 minutes

Plug the value for t into the mini-equation to find the time of the police:

$$t - 1 = 6\frac{2}{3} - 1 = 5\frac{2}{3}$$

The Solution

It took the police $5\frac{2}{3}$ hours or 5.7 hours, rounded to the nearest tenths place, or 5 hours and 40 minutes to catch the bank robbers.

WP25: Two Different Roads, Rates Unknown, Total *Distance* Given

The Problem
Julien commuted 47 miles to work. He drove 45 minutes on the newly paved parkway and 6 minutes through a construction zone. Due to a different speed limit, he drove through the construction zone 40 miles per hour slower than on the newly paved parkway. What were his speeds on the newly paved parkway and the construction zone?

Identify
This is a rate of speed problem in which rates are unknown and the total distance is given (pg 74). There are two unknowns. One will be the variable and one will be in-reference-to the variable.

The Unknowns
Let x = the rate of speed on the paved parkway
Let (x - 40) = the rate of speed through the construction area

The Template
$(t_1)(r_1) + (t_2)(r_2) = d_{total}$

#x + #(x +/- #) = #

The Setup
Since rate of speed is in miles per *hour* units, the given time in *minutes* must be converted to hours. This will facilitate the product of "time" times "rate" to properly yield units of miles, which is the unit of the distance.

Preliminary Conversions of minutes to hours:

$$40 \text{ min} \left(\frac{1 \text{ hr}}{60 \text{ min}} \right) = 0.75 \text{ hr}$$

$$6 \text{ min} \left(\frac{1 \text{ hr}}{60 \text{ min}} \right) = 0.1 \text{ hr}$$

Substitute the times into the template equation to make:
$0.75x + 0.1(x - 40) = 47$

147

The Math
Distribute the "0.1" through the parentheses to get:
$0.75x + 0.1x - 4 = 47$

Move the "-4" to the right by adding it to both sides, then combine like-terms to get:
$0.85x = 51$

Divide both sides by coefficient 0.85:
$$\frac{0.85x}{0.85} = \frac{51}{0.85}$$

$x = 60$

Plug 60 in for x in:
The speed through the construction area $= x - 40 = 60 - 40 = 20$

The Solutions
Julien traveled at 60 miles per hour on the paved parkway and 20 miles per hour through the construction area.

WP26: Two Different Roads, Rates Unknown, Total *Time* Given

The Problem
Carmella took a trip to visit her sister. To get there, she drove 20 miles through the country and 8 miles through the city. She drove at a speed 10 miles per hour slower on the city road than the country road. The total travelling time on both roads was 36 minutes. Before the trip, she put $35 of gas in the car. At what speed did she drive through the country and the city, respectively?

Identify
This is a rate of speed problem in which rates on each type of road are unknown (pg 75). It is also a one-variable problem with two unknowns. One unknown will be the variable and the other unknown is in-reference-to the variable. Since the rate of speed, r, is the unknown, you should expect that the values for the other variables d and t from the rate of speed formula are given and will be used somehow in the setup. Also, since an additional question asks for the time on each road, you must first find the solutions to the first question (find the rates) to use to answer the second question (the times spent on each road). The amount of money spent on gas plays no role in the equation or solving process, so simply disregard it.

Because the variable will be in the denominator, this will setup a rational equation. Find the extraneous solutions, then use the LCD to eliminate the denominators, which will turn it into a quadratic equation.

The Unknowns
Let x = speed in miles per hour on the country road
Let (x – 10) = speed in miles per hour on the city road

Note: You could also use "r" instead of "x".

The Template

$$\frac{d_{\text{one type of road}}}{x} + \frac{d_{\text{another type of road}}}{x \pm \#} = \text{total travelling time}$$

$$\frac{\#}{x} + \frac{\#}{x \pm \#} = \#$$

The Setup

Since the speeds are in units of miles per hour, you must **convert** the total time of 36 minutes to hours to keep the units consistent:

$$36 \text{ min} \left(\frac{1 \text{ hr}}{60 \text{ min}} \right) = 0.6 \text{ hr}$$

Plug that in for "total time" in the template equation, which sets up a rational equation:

$$\frac{20}{x} + \frac{8}{x - 10} = 0.6$$

The Math

Since this is a rational equation, start by finding the extraneous solutions. Set each denominator equal to zero and solve for x.
x = 0, and

x − 10 = 0, so x = 10

We now know that x cannot be 0 or 10. Logically, x couldn't be 0 anyway because driving 0 miles per hours is not moving. We will compare our purported solutions to these near the end of the problem.

Find the LCD, which is "[x(x - 10)]", and multiply it by each term in the equation, which will eliminate all denominators:

$$\frac{20[x(x - 10)]}{x} + \frac{8[x(x - 10)]}{(x - 10)} = 0.6[x(x - 10)]$$

and bring the variable x up to the numerator. In the left-most rational expression, "x" cancels with "x". In the next rational expression, "(x − 10)" cancels with "(x − 10)". On the right side, there's nothing to cancel; distribute the x through "(x − 10)". This makes:
$20(x - 10) + 8x = 0.6[x^2 - 10x]$

Distribute where you can. This becomes:
$20x - 200 + 8x = 0.6x^2 - 6x$

Continued on the next page…

$20x - 200 + 8x = 0.6x^2 - 6x$

Combine like-terms where you can, which, here, are the $20x + 8x$ on the left:
$28x - 200 = 0.6x^2 - 6x$

Since there is a squared variable, move all terms to the right side by subtracting "$28x$" from both sides and adding "200" to both sides. Combine like-terms and arrange into standard form of a quadratic equation, and it becomes:
$0 = 0.6x^2 - 34x + 200$

Put all terms into descending order if it isn't already; it already is in descending order. Set up the Quadratic Formula:
$$x = \frac{-b \pm \sqrt{b^2 - 4ac}}{2a}$$

Substitute into the Quadratic Formula, where a is 0.6, b is -34, and c is 200:
$$x = \frac{-(-34) \pm \sqrt{(-34)^2 - 4(0.6)(200)}}{2(0.6)}$$

Negative (1) times negative 34 becomes positive 34.
Inside the radical, negative 34 squared is 1156, and 4 times 0.6 times 200 is 480.
In the denominator, 2 times 0.6 is 1.2.
$$x = \frac{34 \pm \sqrt{1156 - 480}}{1.2}$$

Inside the radical, 1156 minus 480 is 676
$$x = \frac{34 \pm \sqrt{676}}{1.2}$$

Take the square root of 676, which is 26
$$x = \frac{34 \pm 26}{1.2}$$

Break the "plus or minus" into separate addition and subtraction equations:

$$+: x = \frac{34 + 26}{1.2} = \frac{60}{1.2} = 50$$

and

$$-: x = \frac{34 - 26}{1.2} = \frac{8}{1.2} = 6.7$$

Purported solutions: $x = 50$ and 6.7
Neither of these conflict with the extraneous solutions 0 and 10, found earlier.
How do you know which of the two solutions are valid?
Think about each purported solution logically:

The solution can't be 6.7 because the speed through the city is 10 miles per hour less (slower) than the speed through the country, so 6.7 - 10 would yield a negative speed, which wouldn't be logical. This allows you to conclude that the only solution of x is:

$x = 50$ miles per hour. You must still find the other unknown using this value...

Substitute the 10 in for x to find the speed on the country road:
50 - 10 = 40 miles per hour

The Solutions
Speed on country road = 50 mi/hr
Speed on city road = 40 mi/hr

One More Thing
Suppose the original problem asked an additional question:
How many minutes did Carmella spend travelling on each type of road?

Minutes are units of time so you will solve for t. The units for time in the equation are currently in hours, so solve for time in hours (found earlier), then convert to minutes.

Use each speed to solve for the times on each road using the Rate of Speed Formula:
$$r = \frac{d}{t}$$

Continued on the next page...

For the country road:
$$\frac{50 \text{ mi}}{\text{hr}} = \frac{20}{t_{\text{country}}}$$

Cross multiply to get:
$50t_{\text{country}} = 20$

Divide both sides by 50, to get:
$t_{\text{country}} = 0.4$ hr

Convert from hours to minutes:
t_{city}: $0.4 \text{ hr} \left(\frac{60 \text{ min}}{1 \text{ hr}} \right) = 24$ minutes

Now for the city road:
$$\frac{40 \text{ mi}}{\text{hr}} = \frac{8}{t_{\text{city}}}$$

Cross multiply to get:
$40t_{\text{city}} = 8$

Divide both sides by 40, and

$t_{\text{city}} = 0.2$ hr

Convert from hours to minutes:
t_{city}: $0.2 \text{ hr} \left(\frac{60 \text{ min}}{1 \text{ hr}} \right) = 12$ minutes

Carmella traveled 24 minutes on the country road and 12 minutes on the city road.

WP27: Rate of Speed Unknown but Same; Distances Given, Time Not Given but Reference to Time Given

The Problem
A paperboy rides his bike to deliver papers on two routes, peddling at the same average rate of speed. The first route is one mile and the second is eleven miles. If he does the first route in an hour and a quarter less than the second, what is his average rate of speed?

Identify
This is a rate of speed problem where rate of speed is unknown but the same on each route (pg 76). There is only one unknown, so that will be the variable. Distances are given. Times are not given, nor are they the same, but a reference (compensation number) is given from one time to the other. This will set up a rational equation and will begin to be solved using the LCD.

The Unknown
Let x = the average rate of speed of the paperboy

The Template
Time$_{route\ 1}$ = time$_{route\ 2}$ +/- time compensation #

$$\frac{distance_1}{rate} = \frac{distance_2}{rate} \pm time\ compensation\frac{\#}{\#}$$

$$\frac{\#}{x} = \frac{\#}{x} \pm \frac{\#}{\#}$$

The Setup
Preliminary Step: Since the difference in time is an hour and a quarter, convert it into an improper fraction,

$$1\frac{1}{4} = \frac{5}{4}$$

(you could also convert it into decimal form, 1.25), then plug it in for the time compensation number:

$$\frac{1}{x} = \frac{11}{x} - \frac{5}{4}$$

The Math
The extraneous solution here is "0". Multiply all terms by the LCD which is 4x (continued on the next page):

$$\frac{1(4x)}{x} = \frac{11(4x)}{x} - \frac{5(4x)}{4}$$

Cancel out common factors in the numerators and denominators, eliminating all denominators, then multiply any remaining factors to get:
4 = 44 - 5x

Move the "44" to the left by subtracting it from both sides, then combine like-terms to get:
-40 = -5x

Divide both sides by coefficient "-5":
$$\frac{-40}{-5} = \frac{-5x}{-5}$$

x = 8

The Solution
The paperboy's average rate of speed is 8 miles per hour.

Follow-Up Question
How long does it take the paperboy to deliver papers on each route?

Let $\frac{1}{x}$ = the time to deliver on route 1

Let $\frac{11}{x} - \frac{5}{4}$ = the time to deliver on route 2

Substitute 8 in for x in each:

Time to deliver to route 1 $= \frac{1}{x} = \frac{1}{8}$ or 0.125 hr

Time to deliver to route 2 $= \frac{11}{8} = 1.375$ hr

If you convert the decimal-portion of the hours to minutes:

Route 1: 0.125 hr $\left(\frac{60\ min}{1\ hr}\right) = 7.5$ minutes

Route 1: 0.375 hr $\left(\frac{60\ min}{1\ hr}\right) = 22.5$ minutes

you get: 7 minutes and 30 seconds for route 1 and
1 hour, 22 minutes and 30 seconds for route 2.

WP28: Rate of Vehicle Unknown; Distance & Rate of Current Given; Same Time

The Problem
The Venice Canal moves at an average speed of 3 kilometers per hour. If it takes a couple on a gondola ride the same time to go 2 kilometers upstream as it does to go 8 miles downstream, what would the rate of the gondola be in still water? What is the rate of the gondola moving upstream? What is the rate of the gondola moving downstream?

Identify
This is a rate of speed problem in which rate of speed of the vehicle (gondola) in still water (no current) is unknown (pg 77). The speeds of the vehicle moving upstream and downstream are also unknown but since the speed of the current is given, and the gondola and the speed of the current (in both directions) are in reference to each other, the speed of the gondola will be the variable. Distances are given. Times are not given but they are said to be the same; this is why your set up equation will be built from setting the times (travelling in each direction) equal to each other. This will set up a rational equation. Start by finding the extraneous solutions, then proceed by either cross multiplying or multiplying both sides by the LCD.

The Unknowns
Let x = the rate of the gondola in still water
Let x + 3 = the rate of the gondola moving downstream
Let x − 3 = the rate of the gondola moving upstream

The Template
$$\left(\frac{d}{r}\right)_{downstream} = \left(\frac{d}{r}\right)_{upstream}$$

$$\frac{\#}{x + \#} = \frac{\#}{x - \#}$$

The Setup
$$\frac{8}{x + 3} = \frac{2}{x - 3}$$

The Math
The extraneous solutions are "-3" and "3". Cross multiply to get:
$$8x - 24 = 2x + 6$$

Continued on the next page...

The Math

$8x - 24 = 2x + 6$

Note: You could have also multiplied both sides by the LCD "$(x + 3)(x - 3)$", and, after canceling common factors and distributing, arrived at the same equation.

Move the 2x to the left by subtracting it from both sides, move the "-24" to the right by adding 24 to both sides, then combine like-terms to get:
$6x = 30$

Divide both sides by coefficient 6:
$$\frac{6x}{6} = \frac{30}{6}$$

$x = 5$

Substitute 5 in for x into:
The rate of the gondola moving downstream $= x + 3 = 5 + 3 = 8$
The rate of the gondola moving upstream $= x - 3 = 5 - 3 = 2$

The Solutions
The rate of the gondola in still water is 5 km/hr,
the rate of the gondola moving downstream is 8 km/hr, and
the rate of the gondola moving upstream is 2 km/hr.

WP29: Rate of Vehicle & Current Unknown: Upstream/Downstream (2 Variables)

The Problem
Mark flew 2100 miles from Philadelphia to Las Vegas to attend a bachelor party, and then flew back 3 days later. Flying westward against the jet stream, the flight took 300 minutes. Flying home, eastward, with the jet stream, the flight took 270 minutes. Based on this information, what was the speed of the plane flying west, against the jet stream? What was the speed of the plane flying east, with the jet stream? What was the speed of the jet stream? And what would the speed of the plane be in still-air (no jet stream/air current)?

Identify
This is a problem in which the rates of speed of the vehicle and current are unknown (pg 78). Distance is given and the two total travelling times are given. We will solve using two variables and two equations, but there are four unknowns. We will name two of the unknowns using different arrangements of the two variables (the other two unknowns are just each variable). We will solve using a system of two linear equations.

The Unknowns
Let x = the speed of the plane in still air (no jet stream)
Let y = the speed of the jet stream
Let (x + y) = the speed of the plane *with* the jet stream
Let (x − y) = the speed of the plane *against* the jet stream

The Template
The template equations are built from the rearranged rate of speed formula:
$(t)(r) = d$:

$$\#(x + y) = \#$$
$$\#(x - y) = \#$$

Setup
Put the given values into their associated equations:
1) $4(x + y) = 2400$
2) $6(x - y) = 2400$

Continued on the next page...

The Math

Divide both sides by the coefficient in front of the parentheses:

1) $\dfrac{4(x+y)}{4} = \dfrac{2400}{4}$

2) $\dfrac{6(x-y)}{6} = \dfrac{2400}{6}$

This will convert each equation into:
1) $x + y = 600$
2) $x - y = 400$

This is a system of two linear equations. Let's use the Substitution Method, solving equation 1 for x by subtracting y from both sides, to get:

$x = 600 - y$

Substitute "600 – y" in for x into equation 2, as:
2) $(600 - y) - y = 400$

Simplify by combining like-terms, then subtracting "600" from both sides, to get:

$-2y = -200$

Solve for y by dividing both sides by coefficient "-2":

$\dfrac{-2y}{-2} = \dfrac{-200}{-2}$

$y = 100$

Plug "100" in for y into either of the original equations (or one of the equations after you divided both sides by the coefficient). Let's use *converted equation 1*:
1) $x - 100 = 400$

Add "100" to both sides to get

$x = 500$

Now plug in the values of x and y just found to solve for the other unknowns:

Speed with the jet stream: $500 + 100 = 600$

Speed against the jet stream: $500 - 100 = 400$

The Solutions
The speed of the plane = 500 mi/hr
The speed of the jet stream = 100 mi/hr
The speed of the plane *with* the jet stream = 600 mi/hr
The speed of the plane *against* the jet stream = 400 mi/hr

Note: In "The Math" section, you could have also distributed the coefficient through the parentheses in each equation instead of dividing both sides by the coefficient, and still gotten the same answers. Using distribution would yield larger numbers whereas dividing both sides by the coefficient yields smaller numbers, which is why dividing both sides by the coefficient is encouraged.

See more word problems involving systems of linear equations starting with WP45 (pg 193).

WP30: Travel Times Unknown; References to Rate & Time Given

The Problem
Two friends will be meeting at the Newark Liberty International Airport from which they will depart for vacation. They will each travel exactly 150 miles to the airport, one by car and the other by bus. The car will travel at an average speed of twenty miles per hour faster than the bus, and will take two hours less than the bus to arrive at the airport. How long will it take each of these vehicles to arrive at the airport?

Identify
This is a rate of speed problem where the travel times of each vehicle are unknown. Total distance, which happens to be the same, is given (pg 79). A reference is made to the speed of the bus, and another reference is made to the time of the bus. Notice that rate of speed is not given, and also, there is not enough information to make two equations, so one (rational) equation will be made, setting the rate of the car equal to the rate of the bus, equalized by the given compensation number. Before solving, find the extraneous solution. Then, multiply all terms by the LCD to convert the rational equation to a quadratic equation. Note: Even though distance is the same, you cannot set distance equal to distance in this case, because if you did, due to the given information, both sides would cancel each other out, and x would not be found.

The Unknowns
Let x = the time for the bus to travel
Let x − 2 = the time for the car to travel

The Template
$$\frac{distance}{x} = \frac{distance}{x \pm time \text{ compensation}} \pm speed \text{ compensation}$$

$$\frac{\#}{x} = \frac{\#}{x \pm \#} \pm \#$$

The Setup
$$\frac{150}{x-2} = \frac{150}{x} + 20$$

Note: The compensation number for time was put into the denominator of the left fraction. A compensation number can be put wherever it is needed. The equation could have also been set up as:

$$\frac{150}{x} = \frac{150}{x+2} + 20$$

This setup would also require a re-naming of unknowns as:
Let x = the time of the car, and
Let x + 2 = the time of the bus

Other alternative setups could have the "20" *subtracted* on the left from the left fraction; this would not require the unknowns to be renamed.

The Math
The extraneous solutions are "0" and "-2".
Using the first setup, multiply all terms by the LCD, which is "[x(x – 2)]":

$$\frac{150[x(x-2)]}{(x-2)} = \frac{150[x(x-2)]}{x} + 20[x(x-2)]$$

Cancel out the denominators in both fractions with the appropriate factors in the numerators and distribute the x through the "(x – 2)" on the right to get:
$$150x = 150(x-2) + 20(x^2 - 2x)$$

Distribute where appropriate on both sides to get:
$$150x = 150x - 300 + 20x^2 - 40x$$

Move all terms to the right side, the side with the $20x^2$ (to set up a standard form quadratic equation), by subtracting 150x from both sides, then combine like terms to get:
$$0 = 20x^2 - 40x - 300$$

Divide all terms by the GCF, 20, to make the numbers smaller. The equation becomes:
$$0 = x^2 - 2x - 15$$

Continued on the next page...

$0 = x^2 - 2x - 15$

Attempt to factor into two binomials:
$(x - 5)(x + 3) = 0$

Setting each binomial factor equal to zero and solving yields:
$x = 5$ and $x = -3$

We reject the negative value and choose $x = 5$
Substitute 5 into:
the time for the car to travel $= x - 2 = 5 - 2 = 3$

The Solutions
It takes the person travelling by car 3 hours and takes the person travelling by bus 5 hours.

Note: An additional word problem involving unknown distances travelled by a vehicle on two different roads using gas mileage (miles/gallon) solved with two variables and two equations can be seen in WP51 (pg 205).

SPLITTING A TASK

WP31: Time to Complete Task *Together* Unknown

The Problem

It takes Alison 3 hours to complete a task if she works alone, and it takes Claire 4 hours to complete the same task if she works alone. How long will it take to complete the task if they work together?

Identify

This is a *splitting a task* problem (pg 80) in which the time to complete the task if working together is the only unknown, represented by x. Even though there is one unknown, x, it appears twice (in the numerators of both fractions) in the equation, but you will only get one answer. This sets up a rational equation but since there will be no variables in the denominator, there will not be any extraneous solutions to find. The type of math involved with be:

- Using the LCD to eliminate the denominators. This will set up a 1st degree equation.
- Then: Combine like-terms and solve for x.

The Unknown

Let x = time it takes them to complete the task, working together.

The Template

$$\frac{x}{\text{Alison's time working alone}} + \frac{x}{\text{Claire's time working alone}} = 1$$

$$\frac{x}{\#} + \frac{x}{\#} = 1$$

The Setup

$$\frac{x}{3} + \frac{x}{4} = 1$$

Multiply all terms by the LCD, which is 12:

$$\frac{(12)x}{3} + \frac{(12)x}{4} = 1(12)$$

Continued on the next page...

$$\frac{(12)x}{3} + \frac{(12)x}{4} = 1(12)$$

By canceling out common factors, this will remove all denominators, becoming:
4x + 3x = 12

Combine like-terms on the left to get:
7x = 12

Divide both sides by coefficient 7:
$$\frac{7x}{7} = \frac{12}{7}$$

Rounded to the hundredths place:
x = 1.71 hours
Keep the 1 as an hour and **convert** 0.71 hours to minutes:
$$0.71 \text{ hr} \left(\frac{60 \text{ min}}{1 \text{ hr}}\right) = 43 \text{ min, rounded to the nearest minute}$$

The Solution
It will take them 1.71 hours, or 1 hour and 43 minutes to complete the task, working together.

WP32: Time to Complete Task *Alone* Unknown

The Problem
When Neo and Leo split a job, they can do it together in 3 hours. When Neo does it himself, it takes him 8 hours. How long would it take Leo to do the job if he worked alone?

Identify
This is a *splitting a task* problem (pg 80) where the time for one person to complete the task is the only unknown. Thus, x will represent the only unknown. Also, notice "x" only appears in one place in this type of problem, unlike in WP31 and WP33. This sets up a rational equation. Find the extraneous solutions then proceed by multiplying all terms by the LCD.

The Unknown
Let x = time it takes Leo to do the job alone.

The Template
$$\frac{\text{time together}}{\text{Neo's time}} + \frac{\text{time together}}{\text{Leo's time, x}} = 1$$

$$\frac{\#}{\#} + \frac{\#}{x} = 1$$

The Setup
$$\frac{3}{8} + \frac{3}{x} = 1$$

The Math
The only extraneous solution here would be "0".
Multiply both sides by the LCD, 8x:
$$\frac{3(8x)}{8} + \frac{3(8x)}{x} = 1(8x)$$

Note: Some might ask "could you move the fraction three-eighths to the right, first, then add it with the "1", then cross multiply?" Yes, it can be started in this way as well. However, by starting by multiplying all terms by the LCD allows you to avoid having to add fractions.

Continued on the next page...

$$\frac{3(8x)}{8} + \frac{3(8x)}{x} = 1(8x)$$

By canceling out common factors, this removes all denominators, and becomes:
3x + 24 = 8x

Subtract 3x from both sides to move it to the right side to get:
24 = 11x

Divide both sides by the coefficient 11:
$$\frac{24}{11} = \frac{11x}{11}$$

x = approximately 2.1

The Solution
It would take Leo about 2.1 hours to complete the task if he worked alone. "Approximately" is used because the actual number was rounded to 2.1. The exact form of the answer would be to leave it in fraction form as $\frac{24}{11}$. You could also convert it to decimal form 2.18 (rounded to the hundredths place), or convert the ".18" to minutes as 2 hours and 6 minutes.

WP33: Times for *Each* to Complete Task Unknown

The Problem
When working together, Mario and Angelo can complete three fifths of a task in one hour. It takes Mario forty-five minutes longer to do the whole task alone than it takes Angelo to do the whole task alone. If they can complete the whole task working together in 6 hours and forty-five minutes, how long does it take each of them to complete the task alone?

Identify
This is a *splitting a task* problem (pg 80) in which the times for each to complete the task alone are the unknowns. One unknown will be the variable and the other unknown will be in-reference-to the variable. Start by finding the extraneous solutions, then use the LCD to eliminate the denominators. This will convert the rational equation into a quadratic equation which can be solved either by factoring or by the quadratic formula.

The Unknowns
Since the units of time are in hours, but the compensation time for Mario is given in minutes, **convert** the minutes to hours

$$(45 \text{ min}) \frac{1 \text{ hr}}{60 \text{ min}} = 0.75 \text{ hr}$$

to use in the mini-equation for Mario's unknown time.
Let x = the time it takes Angelo to complete the task alone
Let (x + 0.75) = the time it takes Mario to complete the task alone

The Template
$$\frac{\text{time to complete task together}}{x} + \frac{\text{time to complete task together}}{x \pm \text{compensation time}} = 1$$

$$\frac{\#}{x} + \frac{\#}{x \pm \#} = 1$$

The Setup
$$\frac{1.67}{x} + \frac{1.67}{(x + 0.75)} = 1$$

Continued on the next page...

168

The Setup

$$\frac{1.67}{x} + \frac{1.67}{(x + 0.75)} = 1$$

The Math

The extraneous solutions are "0" and "-0.75". Multiply all terms by the LCD "x(x + 0.75)":

$$\frac{1.67x(x + 0.75)}{x} + \frac{1.67x(x + 0.75)}{(x + 0.75)} = 1x(x + 0.75)$$

Cancel out common factors where appropriate. In the left rational expression, "x" cancels with "x" and in the next rational expression, the binomial factor "(x + 0.75)" cancels with "(x + 0.75). There's nothing to cancel on the right side. Then distribute to get:

$1.67x + 1.2525 + 1.67x = x^2 + 0.75x$

This is a quadratic equation. Round "1.2525" to the nearest hundredths place as "1.25". Combine like-terms on the left side to get:

$3.34x + 1.25 = x^2 + 0.75x$

Move all terms to the right to put the quadratic equation into standard form by subtracting "3.34x" and 1.25 from both sides, then combine like-terms on the right to get:

$0 = x^2 - 2.59x - 1.25$

Use the quadratic formula to solve. Plug in the values and simplify:

$$x = \frac{-(-2.59) \pm \sqrt{(-2.59)^2 - 4(1)(-1.25)}}{2(1)}$$

$$x = \frac{2.59 \pm \sqrt{6.71 + 5}}{2}$$

$$x = \frac{2.59 \pm \sqrt{11.71}}{2}$$

$$x = \frac{2.59 \pm 3.42}{2}$$

Split up the plus and minus into:

$+: x = \dfrac{6.01}{2} = 3.01$

$-: x = \dfrac{-0.83}{2} = -0.42$

Disregard the negative time. Plug "3.01" in for x in:
The time it takes Mario to complete the task alone
= x + 0.75
= 3.01 + 0.75 = 3.76

The Solutions
It takes Angelo 3.01 hours and Mario 3.76 hours to complete the task if they work alone. Converting the decimal portions to minutes would mean it takes Angelo 3 hours and 1 minute (rounded to the nearest whole number) and Mario 3 hours and 45 minutes to complete the task if each works alone.

WP34: PROPORTION & LCD

The Problem
A certain community college keeps a student-to-instructor ratio of 1 to 23. The new semester will have a combined total of 3168 students. How many students and instructors are there in the new semester?

Identify
This is a basic proportion problem (pg 47). Once the rational equation is set up, simplifying can begin either by multiplying the LCD times both sides or by cross multiplying.

The Unknowns
Let x = the number of instructors in the new semester
Let (3168 − x) = the number of students in the new semester

The Template

$$\frac{\text{Instructors}_{\text{standard}}}{\text{Students}_{\text{standard}}} = \frac{x}{\text{Total} - x}$$

$$\frac{\#}{\#} = \frac{x}{\# - x}$$

The Setup

$$\frac{1 \text{ instructor}}{23 \text{ students}} = \frac{x}{3168 - x}$$

The Math
Some textbooks advise you to proceed by multiplying both sides by the LCD in order to eliminate the denominators. Using that approach, the LCD here is "23(3168 − x)".

To get to the same step, you could also *cross multiply* to get:
23x = 3168 - x

Move the "-x" to the left by adding x to both sides, then combine like-terms to get:
24x = 3168

Divide both sides by coefficient 24:

$$\frac{24x}{24} = \frac{3168}{24}$$

x = 132

Plug in 132 for x in:
the number of students in the new semester:
= 3168 – x = 3168 – 132 = 3036

The Solution
There will be 132 instructors to the 3036 students in the new semester.

Note: The extraneous solution is 3168.
For another word problem using proportion and the LCD with triangles, see: WP42 (pg 187).

GEOMETRY

WP35: Perimeter of a Rectangle

The Problem

A rectangle, whose length is 5 inches more than three times its width, has a perimeter of 74 inches. Determine the length and width.

Identify

This references the perimeter of a rectangle (pg 61). There are two unknowns. One will be the variable and one will be in-reference-to the variable. In this specific case, the length is in-reference-to the width.

The Unknowns

If you use L & W (as we will use in this example):
Let W = the width
Let (5W + 3) = the length

If you use x as the variable:
Let x = the width
Let (3x + 5) = the length

The Formula

$2L + 2W = P_{rectangle}$

The Template

$2(\#W +/- \#) + 2W = \#$

The Setup

$2(3W + 5) + 2W = 74$

The Math

Distribute the 2 through the parentheses to get:
$6W + 10 + 2W = 74$

Combine like-terms on the left:
$8W + 10 = 74$

Move the 10 to the right by subtracting it from both sides, to get:
$8W = 64$

Divide both sides by coefficient 8:

$$\frac{8w}{8} = \frac{64}{8}$$

W = 8

Now solve for L by substituting 8 in for W in:
L = 3W + 5 = 3(8) + 5 = 29

The Solutions
W = 8 inches
L = 29 inches

WP36: Perimeter of a Triangle

The Problem
A triangle in which the first side is one cm less than twice the length of the second side, and the third side is one cm more than the second side, has a perimeter of 20 cm. What are the lengths of each side?

Identify
This is a problem involving the perimeter of a triangle (pg 62) where the lengths of all three sides are the unknowns: one of this will be x, and the other two unknowns are in-reference-to x. Notice how the *second side* is *referred-to* by the first and third sides? This is why we will let x be the reference variable to represent the length of the second side. Draw a sketch if it helps you put each side into perspective.

The Unknowns
Let x = the length of the second side
Let (2x - 1) = the length of the first side
Let (x + 1) = the length of the third side

Use **The Formula** for Perimeter of a triangle, in which you add all three sides:

$L_{side\ one} + L_{side\ two} + L_{side\ three} = P_{triangle}$

The Setup
Substitute the unknowns in for each side and plug in the given value of perimeter:

$x + (2x - 1) + (x + 1) = 20$

The Math
The insides of the parentheses cannot be further simplified. Remove the parentheses, taking account of the signs to the left of the parentheses; since these are both plus, it won't change the signs inside.

$x + 2x - 1 + x + 1 = 20$

Combine like-terms. The "-1" and "+1" cancel each other out to zero:

$4x = 20$

Divide both sides by coefficient 4, and

$x = 5$

Plug 5 in for x into the other unknowns to find their lengths.

Side one: $2x - 1 = 2(5) - 1 = 10 - 1 = 9$

Side three: $x + 1 = 5 + 1 = 6$

The Solutions
Side 1 = 9 cm
Side 2 = 5 cm
Side 3 = 6 cm

Note: Problems involving the sum of all angles are solved similar to this one. For a similar problem involving three unknowns with one variable, see WP6 (pg 115).

WP37: Area of a Rectangle

The Problem
A rectangle, whose width is three cm less than four times its length, has an area of 175 cm^2. Determine the length and width.

Identify
This problem involves the area of a rectangle (pg 63) with two unknowns. One unknown will be the variable and the other will be in-reference-to the variable. Here, width is in-reference-to length so L will be the reference variable. Once the unknowns are multiplied, the equation will be revealed to be a quadratic equation.

The Unknowns
Let L = Length
Let (4L − 3) = Width

The Formula
LW = A$_{rectangle}$

The Template
L(#L +/- #) = #

The Setup
L(4L − 3) = 175

The Math
Distribute the L into the parentheses:
$4L^2 - 3L = 175$

Move the 175 to the left side by subtracting it from both sides, in order to put the equation into standard form of a quadratic equation:
$4L^2 - 3L - 175 = 0$

Factor into (4L + 25)(L − 7) = 0

Set each factor equal to 0 and solve. For the left factor, subtract 25 from both sides:
4L = -25

Divide both sides by coefficient 4, and
L = -6.25

For the right factor, add 7 to both sides, and
L = 7

Choose L = 7 because you can't have a negative length (-6.25).

Note: This also could have been found using the Quadratic Formula.

Substitute 7 in for L to find Width in:
Width = 4L – 3 = 4(7) – 3 = 28

The Solutions
The length is 7 cm and the width is 28 cm.

WP38: Area of a Triangle

The Problem
An artist made a triangular poster that she would like to display on a rectangular wall that is 9 ft wide and 12 ft high. The poster has an area of 22 ft^2, and its base is 7 ft shorter than the height. How long are the base and height of the poster? Will the poster fit on the wall without hanging over?

Identify
This is a geometry problem involving the area of a triangle (pg 64). It has two unknowns. One unknown will be the variable and the other will be in-reference-to the variable. This will make a quadratic formula which will need to be solved either by *factor & solve* or by the *quadratic formula*.

The Unknowns
Let h = height
Let the base, b = h − 7

Note, some textbooks use "a" for altitude in place of "h" for height.

The Formula
$$\frac{1}{2}\text{bh} = A_{\text{triangle}}$$

The Template
$$\frac{1}{2}(\text{h} \pm \text{\#})\text{h} = \text{\#}$$

The Setup
$$\frac{1}{2}(\text{h} - 7)\text{h} = 22$$

Note: The dimensions of the wall are not used in the equation, but they are used for comparative purposes at the end, to help answer the last question.

179

The Math

Multiply both sides by the LCD, 2:

$$\frac{(2)1}{2}(h-7)h = 22(2)$$

to eliminate the denominator 2, making:

$(h-7)h = 44$

Distribute the h through the parentheses to get:

$h^2 - 7h = 44$

Subtract 44 from both sides to move it to the right, putting this quadratic equation into standard form:

$h^2 - 7h - 44 = 0$

Attempt for factor into two binomials:

$(h-11)(h+4) = 0$

Set each factor equal to zero and solve to get:

$h = 11$ and -4

Disregard the "-4" because you can't have a negative length. Substitute the "11" into the mini-equation for base:

$b = h - 7 = 11 - 7 = 4$

The Solutions

The height is 11 ft and the base is 4 ft. Since each of these dimensions is less than the respective height and base of the wall, the triangular poster will fit.

WP39: Areas of Two Squares

The Problem
Michele painted a square picture and wants to frame it in a larger, square frame. If a side of the frame is 3 cm less than double the size of a side of the painting, and the area of the painting is 24 cm^2 less than the whole area inside the frame, what are the dimensions of the painting and the frame? What is the area of the painting, what is the area of the whole framed painting? Also, what is the area of just the frame?

Identify
This is a problem involving the areas of two squares (pg 65). The last question is an *area of the shaded region* problem (pg 68). There are two unknowns. One will be a variable and one will be in-reference-to that variable. This will set up a quadratic equation. Also, there will be a binomial squared.

The Unknowns
Let x = the length of a side of the painting
Let (2x − 3) = the length of a side of the frame
Let the area of the painting = x^2
Let the area of the framed painting = $(2x − 3)^2$
Let the area of just the frame = $A_{\text{framed painting}}$ - A_{painting}

The Formula
$A_{\text{square}} = \text{side}^2$

The Template
$(\text{side}_1)^2 = (\text{side}_2)^2$ +/- compensation #

$\#^2 = \#^2$ +/- #
The Setup
$x^2 = (2x − 3)^2$ - 24

The Math
On the right, expand (multiply) the binomial squared to get:
$x^2 = 4x^2$ - 12x + 9 − 24

Move the "x^2" to the right by subtracting it from both sides, then combine like-terms, making a quadratic equation in standard form:
$0 = 3x^2 − 12x − 15$

Attempt to factor; you can factor out a GCF of 3 first:
$0 = 3(x^2 - 4x - 5)$

The quadratic expression in parentheses can then be factored into the binomial factors:
$3(x - 5)(x + 1) = 0$

The coefficient 3 can be divided by both sides causing it to be eliminated. Now set each factor equal to 0 and solve for x to get:

x = 5 and -1, but disregard the "-1" because you can't have a negative length.

Plug 5 in for x in:
The side of the frame = $2x - 3 = 2(5) - 3 = 7$
The area of the painting = $x^2 = 5^2 = 25$
The area of the framed painting = $(2x - 3)^2 = (2(5) - 3)^2 = 7^2 = 49$

The area of just the frame is found by subtracting the area of the painting from the area of the framed painting:
$49 - 25 = 24$
See another problem like this in WP43: Finding the Area of the Shaded Region (pg 189).

The Solutions
Each side of the painting is 5 cm.
Each side of the frame is 7 cm.
The area of the painting is 25 cm^2.
The area of the framed picture is 49 cm^2.
The area of just the frame is 24 cm^2.

WP40: Perimeter of a Triangle & The Pythagorean Theorem

The Problem
A person has just inherited a piece of land which is in the shape of a right triangle. The perimeter was found to be 13.83 km. If one leg is 2 km longer than the other leg, and the hypotenuse is also unknown, what are the lengths of each side?

Identify
This problem deals with the perimeter of a right triangle and, since a right triangle was mentioned, the *Pythagorean Theorem* will be used (pg 67). Note: Most word problems may not blatantly say "hypotenuse," but if a right triangle is involved, it will always have two legs and a hypotenuse. This will ultimately setup a quadratic equation, which can sometimes be solved by first factoring, but can always be solved using the Quadratic Formula.

The Unknowns
Let x = the length of side a (leg 1)
Let x + 2 = the length of side b (leg 2)
Let c = the length of side c, the hypotenuse

The **equation** for this problem is built from a rearrangement of the formula of *perimeter of a triangle* to solve for "c" (in terms of x), which is then substituted in for side c in the *Pythagorean Theorem*:
$a^2 + b^2 = (P - a - b)^2$

See the Detailed Explanation (pg 67) for more background on how this equation is built.

The Setup & The Math
Plug the unknowns for sides a and b, and the value of P into the equation for perimeter:
P: $x + (x + 2) + c = 13.83$

Move all terms except "c" to the right by subtracting them from both sides, then combine like-terms. This *solves for c in terms of x*:

c = -2x + 11.83

Now substitute "-2x + 11.83" in for c into the Pythagorean Theorem. Also, substitute the other unknowns in for a and b, as named earlier. Remember to square each of these:

$$x^2 + (x + 2)^2 = (-2x + 11.83)^2$$

Multiply out the two binomials squared to get:
$$x^2 + x^2 + 4x + 4 = 4x^2 - 47.32x + 139.95$$

Move all terms to the right by the appropriate subtraction on both sides, and combine like-terms, which will put the equation in standard form and is revealed to be a quadratic equation:
$$0 = 2x^2 - 51.32x + 135.95$$

You should attempt to factor into two binomials but since you can't here, let's use the Quadratic Formula. Plug in and simplify:

$$x = \frac{-(-51.32) \pm \sqrt{(-51.32)^2 - 4(2)(135.95)}}{2(2)}$$

$$x = \frac{+51.32 \pm \sqrt{2633.74 - 1087.6}}{4}$$

$$x = \frac{51.32 \pm \sqrt{1546.14}}{4}$$

$$x = \frac{51.32 \pm 39.32}{4}$$

Split up the plus and minus, and simplify, to get:
+: x = 22.66
and
−: x = 3

We reject 22.66 because a side can't be larger than the perimeter (13.83).
The length of *side a* (leg 1) = x = 3

Substitute 3 in for x in:
The length of *side b* (leg 2) = x + 2 = 3 + 2 = 5

Continued on the next page...

$x = 3$

Substitute 3 in for x in:
The length of *side b* (leg 2) = x + 2 = 3 + 2 = 5

Substitute the values for a and b into the perimeter formula:
P: a + b + c = 13.83

= 3 + 5 + c = 13.83

and solve for c. Subtract 8 from both sides to get:
c = 5.83

The Solutions
Leg 1 (side a) is 3 km,
leg 2 (side b) is 5 km, and
the hypotenuse (side c) is 5.83 km.

WP41: Areas of Two Circles

The Problem
A new street will need two new, different sized manhole covers. One
manhole cover will have a diameter of 2 ft and will be 1.25π ft^2 smaller
than the area of the other manhole cover. What is the radius of the larger
manhole cover?

Identify
This problem involves the areas of two circles (pg 66). Notice that the
diameter is given, but radius is part of the formula for Area$_{circle}$, and
radius is also what you are asked to solve for. You can use the diameter
to find radius. This will set up a quadratic equation for which you can
use the square root property (take the square root of both sides) to solve.

The Unknowns
Since radius already goes by variable "r", you don't really have to name
it, but you can acknowledge that r of the larger manhole (circle 2) is the
unknown, which you are solving for.

Let r = the radius of the larger manhole cover (circle)

The Template

$A_{circle\ 1} = A_{circle\ 2}$ +/- Area compensation #

$$\pi(\#)^2_{smaller} = \pi r^2_{larger} +/- \#$$

The Setup

Convert the given diameter, 2, to radius by dividing it by 2, which is 1, so it can be substituted in for r of circle 1:

$$\pi(1)^2 = \pi r^2 - 1.25\pi$$

The Math

$$\pi = \pi r^2 - 1.25\pi$$

The terms with π (without r^2) are like-terms. Combine the like-terms. Add 1.25π to both sides, then combine like-terms on the left of

$$\pi + 1.25\pi = \pi r^2 \text{, to get:}$$

$$2.25\pi = \pi r^2$$

Divide both sides by π which will isolate r^2:

$$\frac{2.25\pi}{\pi} = \frac{\pi r^2}{\pi}$$

$$2.25 = r^2$$

Take the square root of both sides:

$$\sqrt{2.25} = \sqrt{r^2}$$

$$r = +/- 1.5$$

Choose the positive form since you can't have a negative radius.

The Solution

The radius of the larger manhole cover is 1.5 ft

WP42: Similar Right Triangles, Cast Shadow, Proportion/LCD

The Problem
A lamppost with a light at the very top illuminates a street at night and will cast a shadow by anyone who fully stands in its light. A man is standing 10 feet from the lamppost. If his shadow is cast 4 feet in front of him, and if the lamppost is 15 feet taller than the man, how tall is the man and how tall is the lamppost?

Identify
This is a proportion problem involving similar right triangles (pg 68). There will be one (rational) equation and there are two unknowns so one unknown will be the variable and the other will be in-reference-to the variable. Start by finding the extraneous solutions. Since there is a proportion, solving will involve either cross multiplying or multiplying by the LCD.

The Unknowns
Sketch a picture if it helps you put everything into perspective. You want to be able to clearly see the two similar right triangles: the smaller within the larger.

Let x = the difference in height between the man and the lamppost
Let x + 15 = the height of the lamppost

The Template
$$\frac{\text{Base}_{\text{larger}}}{\text{Height}_{\text{larger}}} = \frac{\text{Base}_{\text{smaller}}}{\text{Height}_{\text{smaller}}}$$

$$\frac{\#}{x + \#} = \frac{\#}{x}$$

The Setup
Since we don't yet know the value of the larger base, you can do a preliminary mini-calculation to find the base of the larger triangle made from the tip of the shadow to the base of the lamppost by adding the distance of the shadow in front of the man with the distance between the man and the lamppost.

187

Base$_{\text{larger triangle}}$ = 4 + 10 = 14

Plug this and the other given values into the template equation as:
$$\frac{14}{x + 15} = \frac{4}{x}$$

The Math
The extraneous solutions are "0" and "-15".

You could cross multiply, but let's multiply both sides by LCD "x(x + 15)":

$$\frac{[x(x + 15)]14}{(x + 15)} = \frac{4[x(x + 15)]}{x}$$

On the left side, the factors "(x + 15)" cancels out, and on the right side, the "x" *outside* the parentheses cancels with the "x" in the denominator. Multiply the factors that remain:
[x]14 = 4[x + 15]

to get:
14x = 4x + 60

Move the 4x to the left by subtracting it from both sides. Then combine like-terms to get:
10x = 60

Divide both sides by coefficient 10:
$$\frac{10x}{10} = \frac{60}{10}$$

x = 6

Plug 6 in for x into:
The height of the building = x + 15 = 6 + 15 = 21

The Solutions
The man is 6 feet tall and the lamppost is 21 feet tall. There is no conflict with the extraneous solutions.

WP43: Finding the Area of the Shaded Region

The Problem
Last summer, Roland put up a circular, above-ground pool in his yard. The pool is 21 feet in diameter. Now, he is going to spread fertilizer on his lawn. The fertilizer can be purchased by the square foot. If his yard, which is a perfect 37 ft by 41 ft rectangle, and completely covered with grass except for the pool, how many square feet of fertilizer does he need?

Identify
This is an *area of the shaded region* problem (pg 68). In this problem, the grassy part represents the shaded region and the circular pool represents the cut-out part.

The Unknowns
The only unknown is the area of the shaded region (the grass).
Let A_{grass} = the area of the shaded region

Use the following **Formulas and Strategy**:
$$A_{shaded\ region} = A_{outer\ shape} - A_{innter\ shape(s)}$$

$$A_{outer} = A_{rectangle} = LW$$

$$A_{inner} = A_{circle} = \pi r^2$$

Substitute in for the outer and inner areas into the above equation to build the **template equation**:
$$A_{shaded\ region} = LW_{rectangle} - \pi r^2_{circle}$$

The Setup
Divide the diameter, 21 ft, by 2 to get the radius, 11.5 ft, put "3.14" in for pi, and fill in the givens to get:
$$A_{shaded\ region} = 1517\ ft^2 - (3.14)(11.5\ ft)^2$$

The Math
Compute the arithmetic and solve. Start by squaring "11.5 ft" remembering to square the unit as well, to get:
$$A_{shaded\ region} = 1517\ ft^2 - (3.14)(132.25\ ft^2)$$

$A_{\text{shaded region}} = 1517 \text{ ft}^2 - 415.265 \text{ ft}^2$

$A_{\text{shaded region}} = 1101.735 \text{ ft}^2$, or:

1102 ft^2, rounded to the nearest whole square foot

The Solutions
Roland will need 1102 ft^2 of fertilizer for the lawn.

Note: Another *finding the shaded area* problem involving a picture frame can be seen in WP39 (pg 181).

WP44: Finding the Volume of the Shaded Region

The Problem
A company is contracted to package souvenir glass rectangular-based right pyramids, one per box. They must snuggly pack each one in foam packing that fills the box except for a perfect cutout of the pyramid, which the glass pyramid fits in to prevent from breaking. The box, which is a perfect cube, measures 4 cm in length on each side. The base of the pyramid measures 2 cm in length and 2.5 cm in width, and the height, from the center of the base to the apex, is 3 cm. What volume of foam (in cm^3) is needed per box?

Identify
This is a *volume of the shaded region* problem (pg 68). In this problem, the foam represents the shaded region and the pyramid represents the cut-out part. You will use the formula for volume of a cube and the formula for volume of a right rectangular-based pyramid.

The Unknowns
The only unknown is the volume of the shaded region (the foam).
Let V_{foam} = the volume of the shaded region

Use the following **Formulas & Strategy**

$$V_{shaded\ region} = V_{outer\ shape} - V_{innter\ shape(s)}$$

$$V_{outer} = V_{cube} = (\text{Length of side})^3 \text{ and}$$

$$V_{inner} = V_{\substack{right\ rectangular \\ based\ pyramid}} = \frac{L_{pyr}Wh}{3}$$

Substitute in for the outer and inner volumes into the above equation to build the **equation**:

$$V_{shaded\ region} = L_{cube}^3 - \frac{L_{pyr}Wh}{3}$$

The Setup

$$V_{shaded\ region} = (4\ cm)^3 - \frac{(2\ cm)(2.5\ cm)(3\ cm)}{3}$$

The Math
Compute the arithmetic and solve:
$V_{shaded\ region} = 64\ cm^3 - 5\ cm^3$

$V_{\text{shaded region}} = 59 \text{ cm}^3$

The Solution
The volume of foam needed per box is 59 cm^3

TWO VARIABLES, TWO EQUATIONS

WP45: Two Coins of Different Value (2 Variables)

The Problem
(Using a similar problem as WP5, pg 113): Carl approached a toll-booth which charges $1.25. In his change compartment, he only had nickels and dimes, and had the exact change needed to pay the toll. How many nickels and dimes did he have? (One new piece of information): There are 22 total coins.

Identify
This is a problem involving two different coins of different value. There are two unknowns, and since two totals were given (value and number of coins), this can be solved with two variables and two equations as a system of two linear equations (pg 91).

The Unknowns
Let x = the number of nickels
Let y = the number of dimes

The Template
1) (value of nickel)(# of nickels) + (value of dime)(# of dimes) = total value of all coins
1) 0.##x + 0.##y = #.##

2) # of nickels + # of dimes = total # of coins
2) x + y = #

The Setup
1) $0.05x + 0.10y = 1.25$
2) $x + y = 22$

See Note at the end of the problem.

The Math
The easier approach is to use the Substitution Method. Let's solve for x in equation 2 by subtracting y from both sides to get:
$x = 22 - y$

Substitute "22 – y" in for x into equation 1 as:
1) $0.05(22 - y) + 0.10y = 1.25$

Distribute the "0.10" through the parentheses to get:
$1.1 - 0.05y + 0.10y = 1.25$

Move the "1.1" to the right by subtracting it from both sides, then combine like-terms to get:
$0.05y = 0.15$

Divide both sides by coefficient "0.05":
$$\frac{0.05y}{0.05} = \frac{0.15}{0.05}$$

$y = 3$

Substitute 3 in for y into one of the original equations; let's use equation 2:
$x + 3 = 22$

Subtract 3 from both sides to get:

$x = 19$

The Solutions: (19, 3)
Carl had 3 dimes and 19 nickels. This agrees with the one variable method used in WP5 (pg 113).

Note: From the given information, the equation also could have been setup where
x = the number of dimes and
(22 – x) = the number of nickels,
and solved with one variable and one equation as:
$0.01x + 0.05(22 - x) = 1.25$

WP46: Two Different Priced Tickets (2 Variables)

The Problem
A famous comedian will be returning this year to a local club to perform a new stand-up set. When he performed last year, tickets for front row seats were $60, tickets for regular seats were $30, and brought in a total of $1890. This year, tickets for front row seats cost $65, tickets for the regular seats cost $33, and will bring in a total of $2068. Both performances sold out within an hour that tickets went on sale. Assuming the number of seats and the seating arrangement has not changed from last year to this year, how many seats of each type does the club have?

Identify
This is a problem involving two items (in this case, tickets) at different prices. Two totals are given and there is enough information to make two equations with two variables to be solved as a system of two linear equations (pg 91).

The Unknowns
Let x = the number of front row seat tickets
Let y = the number of regular seat tickets

The Template
Both equations will be built from:
(ticket price)(# of tickets)$_{front\ row}$ + (ticket price)(# of tickets)$_{regular}$ = total revenue

1) #x + #y = #
2) #x + #y = #

The Setup
Previous year: 1) $60x + 30y = 1890$
This year: 2) $65x + 33y = 2068$

The Math
Let's use the *Substitution Method*, and let's solve equation 1 for x. Start by moving the 60y to the right side by subtracting it from both sides. This gives:

1) $30x = 1890 - 60y$

Divide all terms by coefficient 30:

$$\frac{30x}{30} = \frac{1890}{30} - \frac{60y}{30}$$

x = 63 – 2y

Substitute "63 – 2y" in for x into equation 2:
2) 33(63 – 2y) + 65y = 2068

Distribute the 33 through the parentheses:
2079 – 66y + 65y = 2068

Move the 2079 to the right by subtracting it from both sides. Then, combine like-terms to get:
-y = -11

Divide both sides by -1 and

y = 11

Now, substitute "11" in for y in either of the original equations. Let's use equation 1:
1) 30x + 60(11) = 1890

30x + 660 = 1890

Subtract 660 from both sides to get:
30x = 1230

Divide both sides by coefficient 30:

$$\frac{30y}{30} = \frac{1230}{30}$$

y = 41

The Solutions: (11, 41)
The club has 11 front row seats and 41 regular seats.
This agrees with the one variable, one equation method used in WP7 (pg 117).

WP47: Manufacturing Two Different Items (2 Variables)

The Problem
A company manufactures microphones and speakers for smart-phones. Each microphone costs $6 and takes 3 minutes to make. Each speaker costs $7 and takes 5 minutes to make. If a total of $1259 was spent in a total of 772 minutes to make a batch of these parts, how many of each part was made?

Identify
This is a problem involving the manufacturing of two different items. There are two unknowns. Since totals are given for two different units, we can make two separate equations, and therefore can assign two different variables, one for each of the two unknowns. This will make a system of two linear equations (pg 91).

The Unknowns
Let x = the number of microphones manufactured
Let y = the number of speakers manufactured

The Template
1) (cost of part$_1$)x + (cost of part$_2$)y = total cost of combined parts
2) (time spent$_1$)x + (time spent$_2$)y = total time spent on combined parts

1) #x + #y = #
2) #x + #y = #

Note: Depending on the problem, if "total number of parts" is given, you would use:
"x + y = #" instead of one of the equations above.

The Setup
1) $6x + 7y = 1259$
2) $3x + 5y = 772$

The Math
Let's use the Addition/Elimination Method to solve.
Multiply equation 2 through by "-2" which will convert the "3x" into the opposite of the "6x" from Equation 1, so the x-terms can cancel out to zero when the equations are added.

2) $-2[3x + 5y = 772] = -6x - 10y = -1544$

Add the equations:

    1) $6x + 7y = 1259$
New 2) $+ \; -6x - 10y = -1544$
                  $-3y = -285$

Divide both sides by coefficient "-3":

$$\frac{-3y}{-3} = \frac{-285}{-3}$$

$y = 95$

Substitute "95" in for y into either of the original equations. Let's use original equation 2:

2) $3x + 5(95) = 772$

$3x + 475 = 772$

Move the 475 to the right by subtracting it from both sides to get:
$3x = 297$

Divide both sides by coefficient 3:

$$\frac{3x}{3} = \frac{297}{3}$$

$x = 99$

The Solutions: (99, 95)
99 microphones and 95 speakers were manufactured in the batch.

WP48: Mixing Two Chemicals to Make a Final Solution (2 Variables)

The Problem
Nina works in the fragrance industry and was developing a perfume. To get the scent just right, she needed to make a final volume of 2 Liters of a 22% solution made from a mixture of a 3% solution and a 27% solution. What volume of each solution must she use to achieve her desired final solution?

Identify
This is a problem about mixing two chemicals of different percentage concentrations to make a final solution of a defined percentage concentration. Here, the two variables and two equations are used as a system of two linear equations (pg 91) because total volume is given, and that is used to calculate the product of the volume times the percent of the key ingredient for the final solution. In short, two equations can be made using the given volumes and total volume.

The Unknowns
Let x = the volume of 3% solution needed
Let y = the volume of 27% solution needed
Note: the x and y could be assigned to opposite percentages, and you would still get the correct answers.

The Template
1) (% in dec. form)(vol)$_1$ + (% in dec. form)(vol)$_2$ = (% in dec. form)(total vol.)$_{final}$
2) volume + volume = total volume

1) $0.\#\#x + 0.\#\#y = (0.\#\#)(\#)$
2) $x + y = \#$

The Setup
Do the preliminary mini-calculation of multiplying 0.22 times 2 on the right side.
1) $0.03x + 0.27y = 0.22(2)$

1) $0.03x + 0.27y = 0.44$
2) $x + y = 2$

The Math
Since this is a system of two linear equations, solve them as such. It will be easier to use the *substitution method* here. Let's solve for x in the lower equation by subtracting y from both sides to get:
$x = 2 - y$

Substitute "2 – y" in for x into equation 2 as:
$0.03(2 - y) + 0.27y = 0.44$

Distribute on the left to get:
$0.06 - 0.03y + 0.27y = 0.44$

Combine like-terms. This involves moving the 0.06 to the right by adding it to both sides, which gives:
$0.24y = 0.38$

Divide both sides by the coefficient 0.24:
$$\frac{0.24y}{0.24} = \frac{0.38}{0.24}$$

$y = 1.58$, when rounded to the hundredths place.
Substitute 1.58 in for y into either of the original equations. Let's use equation 2:
$x + 1.58 = 2$

Subtract 1.58 from both sides, and

$x = 0.42$

The Solutions: (0.42, 1.58)
The final solution requires
0.42 Liters of 3% solution and
1.58 Liters of 27% solution.

These answers agree with the one variable, one equation method in WP9 (pg 121).

WP49: Investing in Two Simple Interest Investments (2 Variables)

The Problem
Ed received a $5000 rebate from his federal tax return and wanted to invest that money. In order to diversify his portfolio, he put his money in two different investments: part in a simple interest certificate of deposit (CD) and the remaining part in stock in a technology company. In the first year, they earned a combined amount of $462.50 in interest. If the CD had an interest rate of 12% and the stock went up by 7%, how much money did he invest in the CD and how much did he invest in the stock, respectively?

Identify
This problem involves investing two different portions of money into two separate investments, each with their own interest rate. There are two unknowns. Since two totals are given, we can let each of the two unknowns be their own variable and set up two equations. This will be solved as a system of two linear equations (pg 91).

The Unknowns
Let x = the amount of money invested in the CD at 12%
Let y = the amount of money invested in the stock, up 7%

The Template
1) Amount invested in one investment + amount invested in another investment = total amount invested
2) (% in dec. form)($) $_{investment\ 1}$ + (% in dec. form) ($) $_{investment\ 2}$ = total Interest

1) $x + y = \#$
2) $0.\#\#x + 0.\#\#y = \#$

The Setup
Convert each percentage to decimal form by dividing each by 100, which moves the decimal two places to the left.

1) $x + y = 5000$
2) $0.12x + 0.07y = 462.50$

The Math

Let's use the Substitution Method. In equation 1, solve for x by subtracting the y from both sides to get:
$x = 5000 - y$

Substitute "$5000 - y$" in for x in equation 2:
2) $0.12(5000 - y) + 0.07y = 462.50$

Distribute 0.12 through the parentheses to get:
$600 - 0.12y + 0.07y = 462.50$

Move 600 to the right by subtracting it from both sides, then combine like-terms to get:
$-0.05\, y = -137.5$

Divide both sides by coefficient "-0.05":
$$\frac{-0.05y}{-0.05} = \frac{-137.5}{-0.05}$$

$y = 2750$

Substitute 2750 in for y into one of the original equations; let's use equation 1:
1) $x + 2750 = 5000$

Subtract 2750 from both sides, and

$x = 2250$

The Solutions: (2250, 2750)
Ed invested $2250 in the CD and $2750 in the stock.
These answers agree with the other method involving one variable and one equation in WP22 (pg 141).

WP50: Buying Mixed Items at Two Different Unit Prices (2 Variables)

The Problem
Katrina went into the Jersey Shore Boardwalk Candy Shop and spent $21.74 (not including sales tax) on a mixed bag of fudge and salt water taffy. The salt water taffy costs $8.49/lb and fudge costs $11.99/lb. If she bought a total of two and a quarter pounds, how many pounds of each candy did she buy?

Identify
This problem involves buying mixed items at two different unit prices. There are two unknowns. Since two totals (in different units) are given, you can use two variables and make two equations, and solve as a system of two linear equations (pg 91).

The Unknowns
Let x = the weight of salt water taffy purchased at $8.49/lb
Let y = the weight of the fudge purchased at $11.99/lb

The Template
1) pounds$_{fudge}$ + pounds$_{salt\ water\ taffy}$ = total pounds purchased
2) (cost per lb)(lbs)$_{fudge}$ + (cost per lb)(lbs)$_{salt\ water\ taffy}$ = total dollars spent

1) x + y = #
2) #.##x + #.##y = #.##

The Setup
Convert "two and a quarter" to the decimal form number "2.25" and plug it into equation 1.
1) x + y = 2.25
2) 11.99x + 8.49y = 21.74

The Math
Note: If any number comes out with a decimal beyond the hundredths place, round to the nearest hundredths place, as this is the nearest cent.

Let's use the Substitution Method. In equation 1, solve for x by subtracting the y from both sides to get:
x = 2.25 − y

Substitute "2.25 – y" in for x in equation 2:
2) $11.99(2.25 - y) + 8.49y = 21.74$

Distribute 11.99 through the parentheses to get (rounded):
$26.98 - 11.99y + 8.49y = 21.74$

Move the "26.98" to the right side by subtracting it from both sides, then combine like-terms to get:
$-3.5y = -5.24$

Divide both sides by coefficient "-3.5":
$$\frac{-3.5y}{-3.5} = \frac{-5.24}{-3.5}$$

$y = 1.50$ (rounded to the hundredths place)

Substitute 1.50 in for y into one of the original equations; let's use equation 1:
1) $x + 1.50 = 2.25$

Move 1.50 to the right by subtracting it from both sides and

$x = 0.75$

The Solutions: (0.75, 1.50)
Katrina bought 0.75 lb of fudge and 1.50 lb of salt water taffy.
These answers agree with the one variable, one equation method in WP8 (pg 119).

WP51: Distances on Two Different Roads Unknown, Miles/Gallon Given, Total Miles & Total Gallons Given (2 Variables)

The Problem
Erin just purchased a new car which gets 49 miles per gallon on highway roads and 45 miles per gallon on city roads. If, in a week, Erin used a total of 12 gallons of gas for a total of 560 miles, how many miles did she travel on highway roads and how many miles did she travel on city roads?

Identify
This problem involves two variables and two equations and is similar to a rate of speed problem, but in this case, "miles per gallon" (pg 93) is used in place of "miles per hour."

This problem can be solved as a system of two linear equations but can be set up in two ways, either letting the variables represent:
- miles (which is what the question asks for), or
- gallons (in which case you must use the given "miles per gallon" to convert gallons to miles at the end.

We will first look at the setup where the variables represent miles.

The Unknowns
Let x = the miles travelled on highway roads
Let y = the miles travelled on city roads
and

Let $\dfrac{x}{\text{gas rate}_{\text{highway}}}$ = gallons used on highway roads

Let $\dfrac{y}{\text{gas rate}_{\text{city}}}$ = gallons used on city roads

The Template
1) $\text{miles}_{\text{highway}} + \text{miles}_{\text{city}} = \text{total miles}$

2) $\dfrac{x}{\text{gas rate}_{\text{highway}}} + \dfrac{y}{\text{gas rate}_{\text{city}}} = \text{total gallons}$

1) $x + y = \#$

2) $\dfrac{x}{\#} + \dfrac{y}{\#} = \#$

The Setup
1) $x + y = 560$ miles

2) $\dfrac{x}{49} + \dfrac{y}{45} = 12$

Note: See Alternative Setup on page 207.

The Math
Let's proceed using the Substitution Method by solving equation 1 for x by subtracting y from both sides to get:
$x = 560 - y$

Substitute "$560 - y$" in for x into equation 2:
$$\dfrac{(560 - y)}{49} + \dfrac{y}{45} = 12$$

Multiply all terms by the LCD "(45)(49)":
$$\dfrac{(45)(49)(560 - y)}{49} + \dfrac{(45)(49)y}{45} = 12(45)(49)$$

On the left, the 49s cancel out; in the middle, the 45s cancel out, and on the right, nothing can cancel so the numbers must be multiplied to become:

$45(560 - y) + 49y = 26460$

Distribute the 45 through the parentheses to get:
$25200 - 45y + 49y = 26460$

Move 25200 to the right by subtracting it from both sides, then combine like-terms to get:
$4y = 1260$

Divide both sides by coefficient 4 to get

$y = 315$ Continued on the next page…

$y = 315$

Plug 300 in for y into equation 1:
$x + 315 = 560$ miles

and subtract 300 from both sides to get

$x = 245$ miles

The Solution: (245, 315)
Erin drove 245 miles on highway roads and 315 miles on city roads.

ALTERNATIVE SETUP letting x and y represent gallons of gas:
Let x = the gallons of gas used on highway roads
Let y = the gallons of gas used on city roads
Let 49x = the distance travelled on highway roads
Let 45y = the distance travelled on city roads

The Template
1) $\text{gallons}_{\text{highway}} + \text{gallons}_{\text{city}} = $ total gallons

Equation 2 is built around:
$\text{miles}_{\text{highway}} + \text{miles}_{\text{city}} = $ total miles

Each instance of distance as miles on the left of the equation above is made from the following conversion:
$$\frac{\text{miles}}{\text{gallon}}(\text{gallons}) = \text{miles}$$

making equation 2:
2) $\left(\dfrac{\text{mi}}{\text{gal}}\right)(\text{gal})_{\text{hw}} + \left(\dfrac{\text{mi}}{\text{gal}}\right)(\text{gal})_{\text{city}} = $ total miles

1) $x + y = \#$
2) $\#x + \#y = \#$

The Setup
1) $x + y = 12$
2) $49x + 45y = 560$

The Math

This is a system of two linear equations. Let's begin by using the Substitution Method. In equation 1, solve for x by subtracting the y from both sides to get:

$x = 12 - y$

Substitute "12 – y" in for x in equation 2:

2) $49(12 - y) + 45y = 560$

Distribute the 49 through the terms in parentheses to get:

$588 - 49y + 45y = 560$

Move the 588 to the right by subtracting it from both sides, then combine like-terms to get:

$-4y = -28$

Divide both sides by coefficient "-4":

$$\frac{-4y}{-4} = \frac{-28}{-4}$$

$y = 7$ gallons

Substitute 7 in for y into one of the original equations. Let's use equation 1:

1) $x + 7 = 12$

$x = 5$ gallons

Convert gallons to miles using the respective miles-per-gallon for each road:

For y: $(7 \text{ gal}) \dfrac{45 \text{ miles}}{1 \text{ gal}} = 315 \text{ miles}$

For x: $(5 \text{ gal}) \dfrac{49 \text{ miles}}{1 \text{ gal}} = 245 \text{ miles}$

The Solutions: (245, 315)

Erin drove 245 miles on highway roads and 315 miles on city roads. These answers agree with the answers found using the first method. For another Two Variable, Two Equations problem, see: WP29: Rate of Speed Unknown: Upstream/Downstream (pg 158).

THREE VARIABLES, THREE EQUATIONS

WP52: Three Coins of Different Value (3 Variables)

The Problem
Greg has a pocket full of change consisting only of nickels, dimes, and quarters. The total value of the change is $3.35 and there are a total of 25 coins. He has ten more dimes than nickels. How many of each coin does he have?

Identify
This is a three variable problem for which three equations can be set up. Two equations can be built from the totals given (total value and total number of coins). The third equation is built in-reference-to a variable. This will set up a system of three linear equations with three variables and must be solved accordingly. The Detailed Explanations section (pg 94) contains a full step-by-step procedure.

The Unknowns
Let x = number of nickels
Let y = number of dimes
Let z = number of quarters

Note: You could also use n, d, and q.

The Template
1) (value)(# of nickels) + (value)(# of dimes) + (value)(# of quarters) = total value in dollars
2) # of coins + # of coins + # of coins = total # of coins
3) $\text{coins}_{dimes} = \text{coins}_{nickels}$ +/- compensation # of coins

1) $\#x + \#y + \#z$ = total value in dollars
2) $x + y + z$ = total # of coins
3) $y = x$ +/- #

The Setup

For equation 1, convert the value of each coin into dollar (decimal) form and multiply each value times its associated variable.

1) $0.05x + 0.10y + 0.25z = 3.35$
2) $x + y + z = 25$
3) $x + 10 = y$

The Math

Normally, you would need to perform the Addition/Elimination Method on two pairs of equations, making the same variable in each cancel out, leaving you with two new equations in terms of the same two variables. However, since equation 3 is already in terms of only two variables, we only need to cancel out the z-terms from equations 1 & 2 (making new equation 4), then we can find x & y using equations 3 & 4:

In order to set up the z-terms to cancel out, multiply equation 2 through by "-0.25":
2) $-0.25[x + y + z = 25]$
to get:
2) $-0.25x - 0.25y - 0.25z = -6.25$

Add equation 1 and the converted equation 2; line-up like-terms:
1) $0.05x + 0.10y + 0.25z = 3.35$
2) $+ -0.25x - 0.25y - 0.25z = -6.25$
new 4) $-0.2x - 0.15y + \quad 0 \quad = -2.9$

The z-terms have cancelled out. Solve the system of two linear equations (3 & 4) using the Substitution Method (the Addition/Elimination Method could be used as well but substitution will be easier since Equation 2 is already solved for y.
Substitute "x + 10" in for y into the newly made equation 4:
4) $-0.2x - 0.15(x + 10) = -2.9$

Distribute the "-0.15" through the parentheses to get:
$-0.2x - 0.15x - 1.5 = -2.9$

Move the "1.5" to the right by adding it to both sides, then combine like-terms to get:
$-0.35x = -1.4$

Continued on the next page…

-0.35x = -1.4

Divide both sides by coefficient "-0.35":
$$\frac{-0.35x}{-0.35} = \frac{-1.4}{-0.35}$$

x = 4

Substitute 4 in for x in equation 3 to solve for y:
3) y = 4 + 10 = 14

y = 14

Substitute the values of x & y into equation 2 to solve for z:
2) 4 + 14 + z = 25

18 + z = 25

Subtracting 18 from both sides yields:

z = 7

The Solutions: (4, 14, 7)
There are 4 nickels, 14 dimes and 7 quarters in Greg's pocket.

EXPONENTIAL FUNCTIONS

WP53: Exponential (Continuous) Growth

The Problem
If the population in the United States at the end of 1900 was 76 million and grew to 319 million by the end of 2014, what is the growth function that models this data? By this model, what year (rounded to a whole year) was the population in the U.S. three fourths of what it was in 2014? Also, by this model, what was the estimated population in 1970 (rounded to the nearest million)? How close is the estimated population, determined in the previous question, compared to the actual population in the U.S. in 1970, which was about 205 million?

Identify
This is an Exponential Growth/Decay problem, specifically a population growth problem (pg 96).

The Formula
$A = A_0 e^{kt}$

The Setup for Part 1 to find k and the function
At 1900, t = 0

1a. Preliminary calculation to find t at 2014:
$t = \text{year}_{asked} - \text{year}_{initial}$

$t = 2014 - 1900 = 114$

At 2014, t = 114

Plug "114" in for t.
1b. Plug "319" in for A (on the left).
1c. Plug 76 in for A_0.
Leave "e" as "e":

$319 = (76)e^{114k}$

Continued on the next page...

$$319 = (76)e^{114k}$$

The Math for Part 1 to find k and the function

1d. Divide both sides by coefficient 76 in front of e^{114k}:

$$\frac{319}{76} = \frac{(76)e^{114k}}{76}$$

which makes:

$$4.197 = e^{114k}$$

1e. Take the natural log (ln) of both sides:

$$\ln(4.197) = \ln e^{114k}$$

1f. Compute the left side on a calculator to get a number. On the right side, "lne" cancels out leaving:

$$1.434 = 114k$$

Divide both sides by coefficient 114:

$$\frac{1.434}{114} = \frac{114k}{114}$$

to get the value of k:

$$k = 0.0126$$

Now, re-write the formula with "0.0126" in for k, as:

$A = A_0 e^{0.0126t}$ or, as a function: $f(t) = A_0 e^{0.0126t}$

Since A_0 is given and known, you could also write:

$A = 76e^{0.0126t}$ or, as a function: $f(t) = 76e^{0.0126t}$

Use the formula version for Part 2.

The Setup for Part 2 to find t

2a. Begin by finding the value of "A" by doing a mini-preliminary calculation. Since this is a growth question, multiply the fraction times the final amount as:

$$\frac{3}{4}(319) = 76e^{0.0126t}$$

or, if three fourths is converted to the decimal form of 75%:

$$(0.75)(319) = 76e^{0.0126t}$$

213

Either way, the left side will equal 239.25

2b. Plug 239.25 in for A (just found):
$239.25 = 76e^{0.0126t}$

2c. Divide both sides by coefficient 76 in front of e:
$$\frac{239.25}{76} = \frac{76e^{0.0126t}}{76}$$

which will cancel out on the right side and give you a new number on the left as:
$3.148 = e^{0.0126t}$

2d. Take the natural log (ln) of both sides:
$\ln(3.148) = \ln e^{0.0126t}$

2e. Compute the left side on a calculator to get a number. On the right side, "lne" cancels out leaving 0.0126t, as:
$1.147 = 0.0126t$

2f. Divide both sides by coefficient 0.0126:
$$\frac{1.147}{0.0126} = \frac{0.0126t}{0.0126}$$

t = 91 (rounded to the nearest whole year)

2g. Add 91 to the starting year to get the year of the given population:
$1900 + 91 = 1991$

The Setup for Part 3
Preliminary calculation:
Find t for 1970:
$1970 - 1900 = 70$

t = 70

Plug 70 in for t into the function you determined from Part 1:
$f(70) = A = (76)e^{(0.0126)(70)}$

Continued on the next page...

The Math for Part 3

Following order of operations, simplify the exponents of "e" first, as this is technically a group, by multiplying to get:

$A = (76)e^{0.882}$

Since exponents are then next priority by order of operations, compute "$e^{0.882}$" to get:

$A = (76)(2.4157)$

Note: Be sure to use the proper "e" on your calculator; do not use the button for scientific notation.

Multiply to get:

$A = 184$, rounded to the nearest million

Part 4

Subtract the answer from Part 3 with the actual population (given):

$205 - 184 = 21$

The Solutions

1. $k = 0.0126$ and the function/formula is

$f(t) = A = A_0 e^{0.0126t}$ or $f(t) = A = 76e^{0.0126t}$

2. The year the population became three fourths of the population from 2014 was 1991.

3. In 1970, the population is estimated to be 184 million.

4. The population predicted by this model was off (under) the true population of 1970 by 21 million.

WP54: Logistic Growth of a Viral Epidemic (Influenza)

The Problem
An influenza epidemic began on a college campus upon students' return from Thanksgiving break. The collected data made the function:

$$f(t) = \frac{15000}{1 + 20e^{-1.3t}}$$

which can be used to tell how many students had the flu t weeks since the initial outbreak, and to answer the following questions:
1. How many students were infected when the epidemic began?
2. How many students were infected by the end of the third week?
3. At what time was half of the student body infected?
4. At what time did the rate of infection start to slow down? (Answer by giving the point of maximum growth.)
5. What is the limiting size of the population to get infected?

Identify
This is a logistic growth model problem (pg 101), specifically applied to an epidemic. Note that t is in units of weeks. Also, when reporting units of people, the number must be rounded to the nearest whole.

The Formula – All questions will use the given function:
$$f(t) = \frac{15000}{1 + 20e^{-1.3t}}$$

1. The Setup & The Math
Find f(0); plug 0 in for t and simplify:
$$f(0) = \frac{15000}{1 + 20e^{-1.3(0)}}$$

"(-1.3)(0)" equals 0,

$$= \frac{15000}{1 + 20e^{0}}$$

and $e^0 = 1$, as any non-zero base raised to 0 equals 1:
$$= \frac{15000}{1 + 20(1)}$$

Continued on the next page...

$$= \frac{15000}{1 + 20(1)}$$

"(20)(1)" equals 20, so:

$$f(0) = \frac{15000}{1 + 20} = \frac{15000}{21} = 714, \text{rounded to the nearest whole}$$

Note: This could have also been determined by finding the y-intercept.

2. The Setup & The Math
Simplify, following order of operations:
$$f(3) = \frac{15000}{1 + 20e^{-1.3(3)}}$$

Multiply the exponents of e to get:
$$= \frac{15000}{1 + 20e^{-3.9}}$$

Compute $e^{-3.9}$ to get:
$$= \frac{15000}{1 + 1.579}$$

Add the numbers in the denominator to get:
$$= \frac{15000}{2.579}$$

Divide, and
$$f(3) = 5816$$

3. The Setup & The Math
To find half of the student body, divide 15000 by 2 which is 7500 and put this in for A:
$$7500 = \frac{15000}{1 + 20e^{-1.3t}}$$

Put the denominator in parentheses and cross multiply, but do not distribute:

$$7500(1 + 20e^{-1.3t}) = 15000$$

Divide both sides by 7500:
$$\frac{7500\,(1 + 20e^{-1.3t})}{7500} = \frac{15000}{7500}$$

which eliminates 7500 from the top and bottom on the left. On the right, divide to get a new number:
$1 + 20e^{-1.3t} = 2$

Subtract 1 from both sides to get:
$20e^{-1.3t} = 1$

Divide both sides by the coefficient in front of e, which is 20, which isolates the e-term:

$$e^{-1.3t} = \frac{1}{20}$$

On the right, divide to get a number:
$e^{-1.3t} = 0.05$

Take the natural log (ln) of both sides in order to eliminate the "e":
$\ln e^{-1.3t} = \ln(0.05)$

On the left, "lne" is eliminated leaving the exponent, and on the right, compute the number, to get:
$-1.3t = -3$

Divide both sides by coefficient "-1.3" to get:
$t = 2.3$

This could be answered more specifically by multiplying the decimal portion, .3, times 7 days per week:

$$(.3 \text{ weeks})\frac{7 \text{ days}}{1 \text{ week}} = 2.1 \text{ days}$$

to get 2 weeks and 2 days (rounded to the nearest day).
Or, if you would argue that you can't answer as a fraction of a day, you might round up to 2 weeks and 3 days.

Continued on the next page...

4. The Setup & The Math

Since the point where the rate of infection slows down is the point of maximum growth, use the mini-formulas to find the x and y values of the point:

$$x = \frac{\ln(a)}{b} = \frac{\ln(20)}{1.3} = \frac{3.00}{1.3} = 2.308$$

$$y = \frac{15000}{2} = 7500$$

Point of maximum growth: (2.308, 7500)

5. No math is needed to determine the answer. The answer is c, which in this problem is 15000.

The Solutions

1. 714 people were infected when the epidemic began.
2. 5816 people were infected by the end of the third week.
3. Half the student population, 7500 students, were infected at t = 2.3 weeks, or 2 weeks and 2 or 3 days.
4. The rate of infection began to slow at t = 2.308, or as the point of maximum growth: (2.308, 7500)
5. The limiting size of the population to get infected is 15000.

WP55: Compounding Interest

The Problem
You are about to invest $5000 over 30 years and you have two investment options. The Compounding account has an annual interest rate of 7% and is compounded quarterly. The Continuous Growth account (compounding annually) has an annual interest rate of 8%. Which option will result in more money by the end of the 30 years?

Identify
This is an investment problem where you must do two separate calculations with the same given information, then compare the results of each. One will use the Compounding model (pg 104) and the other will use the Continuous Growth model (pg 104 & pg 96).

The Compounding Model
The Formula
$$A = P\left(1 + \frac{r}{n}\right)^{nt}$$

The Setup
$$A = 5000\left(1 + \frac{0.07}{4}\right)^{4(30)}$$

The Math
There are two groups which must be simplified: the exponents and the base to the exponents which is the group in parentheses. Simplify them both, separately:
$$A = 5000(1 + 0.0175)^{120}$$

$$A = 5000(1.0175)^{120}$$

Take care of the exponent by computing 1.0175 raised to 120 to get:
$$A = 5000(8.019)$$

Multiply, and
$$A = 40095.00$$

Note: Depending on how you input the numbers into the calculator, you may get a slightly different answer. If you left the result of "1.0175^{120}" in the calculator, then multiplied by 5000, you will get A = 40095.92, rounded to the hundredths place, (the nearest cent). Although they are slightly different numbers, either is acceptable.

The Continuous Growth Model
The Formula
$A = Pe^{rt}$

The Setup
$A = 5000e^{(0.03)(30)}$

The Math
Simplify the exponents first by multiplying to get:

$A = 5000e^{0.9}$

Compute $e^{0.9}$ on the calculator (and round to the hundredths place) to get:
$A = 5000(2.46)$

Multiply, and you will get
$A = 12,300.00$

Note: If you left the result of $e^{0.9}$ in the calculator and multiplied by 5000, you would get:
$A = 12298.02$, rounded to the hundredths place (nearest cent). Although this yields a slightly different number, either is acceptable.

Comparison of Solutions
The total amount from the Compounding Model at 7% compounding quarterly will be $40,095.00.
The total amount from the Continuous Growth Model at 8% will be $12,300.00.
According to these models, the Compounding account, which compounds quarterly, shows to be the better account to invest in, even though the annual interest rate is 1% less than the Continuous Growth account.

CLOSING

Thank you for using my third algebra study guide book. I hope it reveals the simpler side to word problems and that it helps you excel in your class and your career. If you feel this book helped you, please leave a positive review, please tell your instructors about it, and please recommend it to family and friends. I would also love to hear your feedback so I can get a better idea of what elements helped, which elements didn't help enough, or what topics could have been addressed that were not. You can email me at my personal address: bullockgr@gmail.com.

Also, look for these other books by the author:

ALGEBRA IN WORDS: A Guide of Hints, Strategies and Simple Explanations

ALGEBRA IN WORDS 2: MORE Hints, Strategies and Simple Explanations

ALGEBRA IN WORDS 3: Notes for Algebra 2, College Algebra & Pre-Calculus on Functions, Polynomials, Theorems, Rational Functions & Systems of Equations (Kindle edition)

COLLEGE SUCCESS: An Insider's Guide to Higher GRADES, More MONEY, and Better HEALTH

CPSIA information can be obtained
at www.ICGtesting.com
Printed in the USA
BVHW041918151121
621711BV00007B/64